Gerd Moser

SAP® R/3® Interfacing using BAPIs

Other titles in Computing

The Efficiency of Theorem Proving Strategies
from David A. Plaisted and Yunshan Zhu

Recovery in Parallel Database Systems
from Svein-Olaf Hvasshovd

Applied Pattern Recognition
from Dietrich W. R. Paulus and Joachim Hornegger

Efficient Software Development with DB2 for OS/390
from Jürgen Glag

Corporate Information with SAP®-EIS
from Bernd-Ulrich Kaiser

SAP® R/3® Interfacing using BAPIs
from Gerd Moser

Gerd Moser

SAP® R/3® Interfacing using BAPIs

A practical guide to working within the SAP® Business Framework

Die Deutsche Bibliothek – CIP-Einheitsaufnahme

Moser, Gerd:
SAP R/3 interfacing using BAPIs: a practical guide to working within the SAP business framework / Gerd Moser. – Braunschweig; Wiesbaden: Vieweg, 1999
ISBN 3-528-05694-0

1st Edition 1999

Vieweg is a subsidiary company of Bertelsmann Professional Information.

Printing and binding: Lengericher Druckerei Hubert & Co., Göttingen
Printed on acid-free paper
Printed in Germany

ISBN 3-528-05694-0

Foreword

PricewaterhouseCoopers (www.pwcglobal.com), the world' s largest professional services organisation, helps its clients build value, manage risk and improve their performance. Drawing on the talents of more than 144,000 people in 152 countries, the organisation provides a full range of business advisory services to leading global, national and local companies and to public institutions.

The Management Consulting Services (MCS) practice of PricewaterhouseCoopers helps clients maximise their business performance by integrating strategic change, process improvements and technology solutions. Through a world-wide network of skills and resources, consultants manage complex projects with global capabilities and local expertise, from strategy through implementation. PricewaterhouseCoopers is a recognised leader in the integration of change management and information technology.

Delivering such services to our clients requires a special dedication to developing the technical expertise required to successfully implement such applications as SAP R/3. In establishing SAP Centres of Expertise around the World, PricewaterhouseCoopers has sought to create a premier network of facilities where our professionals can develop and expand their knowledge in leading edge technologies. It was from the European SAP Centre of Expertise in Walldorf, Germany that this publication was created and it is from here, and other Centres around the World, that we continually develop the leading edge skills in new and emerging technologies required by our clients today.

PricewaterhouseCoopers is the world' s largest globally-focused SAP practice and the most successful. With such facilities as the European SAP Centre of Expertise and with the dedication of our professionals, such as Gerd Moser the author of this work, we will continue to deliver first class solutions to our clients, worldwide.

Andrew Gibbs
PricewaterhouseCoopers
European SAP Centre of Expertise
Walldorf, Germany

Contents

Acknowledgements

During the time I was busy with this book, when I was wondering around trying to find information, analysing, structuring, coding, documenting and going almost crazy, I was always thinking about how blessed I am to have parents standing behind me. They always gave me the support, the freedom and the chances I needed to become the man I am now.

On the other side I had my wife, Regina, who supported me all the way through my stress by correcting all the grammar, orthography and expressions and by taking over the duties that I could not comply during this time.

As far as my working environment is concerned, I was also very lucky because I had my Partner Ramon Demelbauer and also my Manager of the ECOE, Roger Arpagaus who generously supported my research wherever they could. Additionally, I want to thank them for enabling the creation of this book.

Very special thanks to all my colleagues who always had an open ear when I needed someone to discuss my problems, my progress and my experience in this field of technology.

In particular, I want to thank the ASEC team in Frankfurt for the system support as well as Marika Engel, Christiane Hachenberger, Thomas Kluge and Stephan Vogt.

Every time I needed to have something discussed in the field of Lotus Notes, Harald Reinartz (Lotus/SAP Competence Centre) was a willing partner and put great effort in helping me.

For providing an intellectual environment that makes working life fun, I would like to thank my Manager of the Technology Centre, Andrew Gibbs for all his good support.

Last but not least, I would like to express my gratitude to Heike Link who read everything and had the courage to tell me when it would not do. She also helped me to maintain the red thread throughout the book.

Without all of them, no book!

Gerd Moser
Walldorf, Germany

Preface

Objectives and main points

The objective of this book is to serve as a tutorial on the technical as well as the business aspects of SAP interfacing technology, with particular emphasis on SAP's newly introduced Business Application Interfaces (BAPIs). The book will attempt to answer questions like:

- What is the concept of SAP's Business Framework Architecture and what is its future role?
- What are BAPIs?
- How can BAPIs be used to improve my application development?
- What are the key issues in programming with BAPIs?
- Which other interface technologies can be used beside the BAPIs?
- What is SAP's strategy for interfacing technologies? What are the technical challenges in connecting an external system with an R/3 system.
- What is the necessary SAP terminology and how do the terms and concepts interrelate?

The reader should be familiar with basic Information Technology (IT) terminology and its concepts.

Structure of the Book

The book is organised into the following chapters:

- **Chapter 1** – gives an overview about the market situation in the groupware and corporate application market. In particular, it looks at the two software companies Lotus and SAP, their flagship products Notes and R/3, and how a combination of them could close existing business gaps.
- **Chapter 2** – Groupware, Workflow and Workflow management are fast evolving technologies. To understand the concepts and how they relate to each other, this chapter defines

and compares them. A description of Lotus Notes, its architecture and its application development environment is covered in the second part of this chapter.

- **Chapter 3** – Corporate Applications, their historical development and their impact on IT, especially with the introduction of the SAP systems are also described. It gives an overview of R/3's functionality and Client/Server architecture and its development environment.
- **Chapter 4** – describes the R/3 system interfaces and how they relate to the ISO/OSI seven-layer model. It gives a short description of each of the interfaces used within a R/3-system.
- **Chapter 5** – This chapter focuses on the advantage of a connection between an R/3 and Lotus Notes system. It then analyses the combination of both systems and the resulting benefits for an enterprise.
- **Chapter 6** – This chapter describes Notes' object-oriented programming language LotusScript. It also describes the built-in extensions of LotusScript, LotusScript extensions (LSX) and their use in developing Notes applications.
- **Chapter 7** – gives an introduction to SAP's Business Framework Architecture (BFA) and its essential parts the Business Components, Business Objects and their interfaces, the BAPIs. Further on it shows how Business Objects are represented in the R/3 system and also how R/3's object-oriented technology can be understood.
- **Chapter 8** – This chapter describes BAPIs and their implementation within the R/3 system. It also takes a closer look at the underlying technology, which is the RFC technology.
- **Chapter 9** – This chapter discusses the technology which stands behind SAP's Application Link Enabling (ALE) and its interrelating technologies. The combination of BAPIs and ALE and their future deployment are discussed in more detail.
- **Chapter 10** – This chapter starts with the motivation for developing a prototype based on Lotus Notes 4.5 and SAP R/3 including a perspective of the aspired goals. It maps the prototype architecture, menu structure and describes each submenu, which will be implemented.

- **Chapter 11** – This chapter describes the technical realisation of the TES prototype. It includes and explains the coding of the various functions, which have been implemented in Lotus Notes and the R/3 System.
- **Chapter 12** – provides a summary of the experiences gained from working with these technologies, theoretically and practically and their future perspectives.

Intended Audience

This book is primarily aimed at information technology (IT) practitioners/managers and information systems students. It can be used in academic courses or in corporate training, or as a self-learning tool and particularly as a reference for SAP terminology.

1 Introduction

As former national markets have become more and more globalised, former national key players are increasingly under pressure to secure their competitive edge. This process of market globalisation and its growing influence on companies' business processes plays a major role on how companies use state of the art information technology (IT). Corporate applications based on client/server technology, for instance, are implemented to integrate business information across the enterprise in real time.

In most cases existing IT systems do not fit the business needs of an organisation. Therefore an IT infrastructure must be built that fully supports the operative business and information sharing across the enterprise. There are two system approaches which can help to close the above-mentioned gap.

The first approach should solve the employees' need of sharing; exchanging and gathering information about projects, customers etc., in other words sharing the collective knowledge, which was in former times contained in a central filing cabinet. One major problem of this was that the mobile employees could not access the data when they were out of the office. Another consequence of this absence was a loss of time with the result that the filed information became out of date.

Groupware, workflow and document management technologies support companies in improving their communication and collaboration between work groups. Their functionality enables users to manage paper documents by routing them electronically through Intranets or the Internet, and controlling computer-generated documents and data through document library functions [PW97a]. Groupware contains a large number of complementary technologies such as e-mail, conferencing, document storage, workflow and because of this it is also regarded as a vague and contested concept [Bro97].

This book focuses on commercial groupware tools, in particular the most successful, commercially available system – Lotus Notes.

Over the last years Lotus Notes has gained a leading market position in the groupware industry. According to a survey conducted by Forrester Research, Inc., one of the world' s leading

IT-analysts, Lotus has nearly 30 percent of the professional groupware market and 73 percent [Forr96] of the messaging market.

Recently Notes has faced growing competition from other products, like Oracle's InterOffice, Microsoft's Exchange, Netscape's Collabra Share and Novell's GroupWise. However a change to one of these applications away from Lotus Notes would be very time consuming and costly (i.e. staff training, software- and development costs), and so Lotus' market share can be seen as very stable. In addition, regarding the new-buyer market Notes has a Unique Selling Preposition (USP) against its competitors in the field of its highly advanced Application Development Environment which is based on huge and sophisticated function libraries.

The second software group should support the business processes of a company in the following business areas: Finance/Accounting, Sales and Distribution, Logistics/Warehousing, Production as well as Human Resources. These functions are covered by Enterprise Resource Planning (ERP) packages which in most cases touch all areas of an organisation.

A major solution provider for this market section is the German company SAP AG (Systems, Applications and Products in Data Processing) with their product R/3 (Real-Time-System, Version 3). With more than 18,000 R/3 installations in over 100 countries, SAP AG is a leading global provider of client/server business application solutions [SAP98].

As described above both SAP and Lotus are leading solution providers in their market section. Therefore a combination of SAP R/3 and Lotus Notes should provide a powerful solution to deliver major business benefits and competitive advantage for today's globalisation challenges.

2 Groupware and related Terminologies

2.1 Groupware and Computer-Supported Co-operative Work

The rapid improvement of information management and the new potentials for communication between people have become essential to the success of most organisations. The increased availability of computer networks and the trend towards teamwork have played major roles.

Research activities in the field of 'computer support for team work' are given the terms *Groupware* or *Computer-Supported Co-operative Work* (CSCW).

What is CSCW?

CSCW is an interdisciplinary research area (see figure 2-1), which is based on computer sciences, sociology, psychology, work and organisational sciences, anthropology, ethnography, business-oriented computer sciences and business studies. Its main emphasis lies on groupwork, supported by information- and communication-technologies [Ste96].

Figure 2-1: The interdisciplinary field of CSCW [TSMB95]

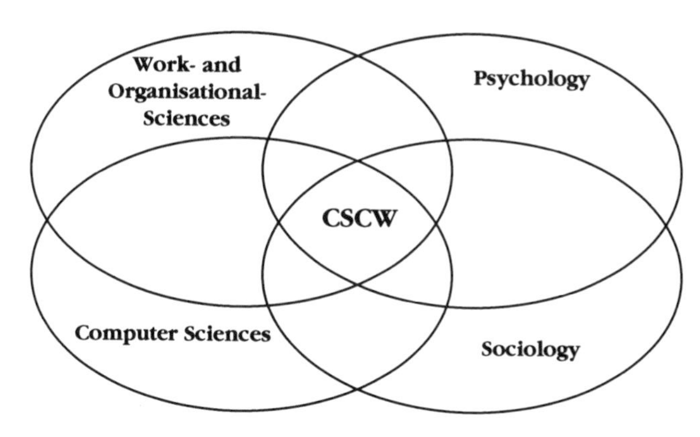

"CSCW is the specific discipline that motivates and validates groupware design. It is the study and theory of how people work together and how the computer and related technologies affect group behaviour" [GS91].

According to Wilson, "CSCW" is "a generic term which combines the understanding of the way people work in groups with the enabling technologies of computer networking, and associated hardware, software, services and techniques" [Wil88].

What is Groupware?

The final result of a transposition of the CSCW's theoretical approach into practice is called groupware. This transposition is achieved taking the CSCW-results into consideration and put into an information- and communication-system that supports teamwork. Groupware is influenced by the user, the task, the organisation and the technology [Ste96].

Ellis defines groupware as "computer-based systems that support groups of people engaged in a common task (or goal) and that provide an interface to a shared environment" [EGR91].

While groupware refers to real computer-based systems, the notion CSCW stands for the study of tools and techniques of groupware as well as their psychological, social and organisational effects.

The main functionality of groupware products is described by the following points [PW96a]:

Graphical User Interfaces (GUIs) – The use of pictures rather than just words to represent the input and output of a program. The program displays certain icons, buttons, dialogue boxes etc. in its windows on the screen, often described as *WIMP* (**W**indows, **I**cons, **M**enus and **P**ointers). The user controls it mainly by moving a pointer on the screen (typically controlled by a mouse, also known by the term *Direct Manipulation (DM)*) and by selecting certain objects through pressing buttons on the mouse while the pointer is pointing at them.

- **On-line Communication** – Discussions can be held in real-time with active interaction between participants.
- **Replication** – Replication is the process of duplicating and updating data in multiple computers on a network; some of the computers have to be permanently connected to it (e. g. servers). Others, such as notebooks or desktops, can be connected occasionally to 'replicate' the changed documents.
- **Remote Access** – Using database replication, remote users can access and update information held in a groupware database.

- **Workflow**[1] – Workflow software moves a *transaction* through the steps required for its completion. In other words, this software is responsible for 'guiding' a transaction from its start to its end.

LUW

ACID-Properties

According to [Date95; DHLW96], a transaction, also known as *Logical Unit of Work* (LUW), is a collection of operations on the physical and abstract application state. Furthermore to satisfy the *Object Management Group*[2] (OMG) transaction service specifications, a transaction should have the following *ACID*-properties: ***A****tomic*, its result should be ***C****onsistent*, ***I****solated* (independent of other transactions) and ***D****urable* (its effect should be permanent).

Groupware sets the basis for possible improvement, as far as productivity is concerned, by providing the following benefits [PW97a]:

- **Communication capability** – Supports communication among project personnel by a variety of tools, including e-mail and discussion forums.
- **Information availability** – As project personnel and project information are linked together, the project information always remains current and accessible, so that the right information gets to the right people at the right time.
- **Location and time independence** – Allows project personnel to be geographically dispersed with no regard to location or time.
- **Project management** – Allows controlled management of project documents, schedules and personnel.

2.2 Workgroup Computing

The term *Workgroup Computing* can be seen as the practical usage of groupware-technology. It is an extension of local networking. When Workgroup Computing is seen in contrast to *Workflow Management*[3], then Workgroup Computing is the method of organising a business around productive teams by using computer support to enable co-operative working and to

[1] Workflow will be described in detail in section 2.3

[2] The Object Management Group, Inc. (OMG) is a consortium of users, software and hardware vendors, which are dedicated to standards for distributed, heterogeneous interoperability, such as CORBA (also see chapter 5). The OMG is also setting standards in object-oriented programming [@OMG].

[3] Workflow Management will be described in section 3.3

eliminate time/space restrictions [Ste96]. One major example of a workgroup computing software is LinkWorks[4] from Digital Equipment Corporation (DEC).

2.3. Workflow and Workflow Management

2.3.1 What is Workflow?

Joosten describes *workflow* as "the movement of documents around an organisation for purposes including sign-off, evaluation, performing activities in a process and co-writing" [Joo94].

The *Workflow Management Coalition*[5] (WfMC) defines Workflow as "the automation of a business process, in whole or part, during which documents, information or tasks are passed from one participant to another for action, according to a set of procedural rules" [Ste96].

In other words, the workflow software manages operational processes performed by an organisation by moving process information through a network, tracking it and providing a functionality which can solve possible network congestions through traffic modelling techniques [PW97a].

2.3.2 Workflow Technologies

Today's workflow systems range from a relatively simple process mapping tool, where the workflow has to be programmed and customised separately, to systems that map and integrate the workflow automatically into the application, without any need for programming [PW96a].

Most workflow companies have split up the workflow-technology into three different bases [PW97a]:

(a) **Production processing** – A highly structured workflow with clearly defined processes, work steps and tasks with hardly vary in daily transactions. For each set of documents, specific work rules are defined (i. e. invoice is handled differently from a quote). The workflow is generally built to

---

[4] LinkWorks is DEC's extensible object-oriented workgroup framework built on open, distributed client/server services supporting integration of custom and third party personal, business and groupware applications into a secure and robust heterogeneous, multivendor environment [@DEC].

[5] Described in detail in section 3.3.3

justify the various business steps of a document, including rules of behaviour when the status of a document changes.

(b) **Enterprise processing** – This type of processing is not as deeply structured as the aforementioned; it only controls the document flow and the work route between the different departments.

(c) **Ad hoc processing** – Roughly similar to an e-mail system, when regarding the definition of routing the documents but with the additional option of applying certain exception conditions ad hoc.

2.3.3 Workflow Management

The definition of the term *Workflow Management* is made by the WfMC[6]. The WfMC is an international organisation which was founded in August 1993 as a non-profit organisation of workflow vendors, users and analysts. Its mission is to promote the use of workflow by establishing standards for software terminology, interoperability and connectivity between workflow products.

The WfMC defines Workflow Management as the modelling, the simulation as well as the execution and the control of business processes supplying the required tools and information. In this case a business process signifies the quantity of independent activities which are dependent on the organisational structure and the management goal [Ste96].

A *Workflow Management System* (WFMS) supports the automation of business processes and co-ordinates the schedule of work items and the activities assigned to the involved people. It also provides the software functions that are essential for the automation of the business processes. Therefore it should support [BG96]:

- Co-ordination of one or more transactions which are assigned to a business process.
- *Worklist* (a list of work items [Holl94]) *Management* for an active administration of available workload which can be assigned to certain employees or workgroups.
- *Event Management* for a flexible reaction to sudden events, such as status-changes of raw material or incoming messages from external systems.

[6] For further information see: [@WFMC]or [TSMB95]

- *Deadline Management* for automatic control of cost- and time-deadlines.
- Optimisations of business processes with the help of modelling and simulation.

Most of today's workflow systems have remained document management systems, which are based on relatively simple, organisational processes but not on integrational complex business steps. Despite this, there are already vendors, who have incorporated the above mentioned tasks into their systems, e.g. SAP.

Major representatives of WFMS systems include FlowMark[7], InConcert[8] and SAP's Business Workflow[9].

Figure 2-2.: Workflow Reference Model-Components & Interfaces [Holl94; TSMB95]

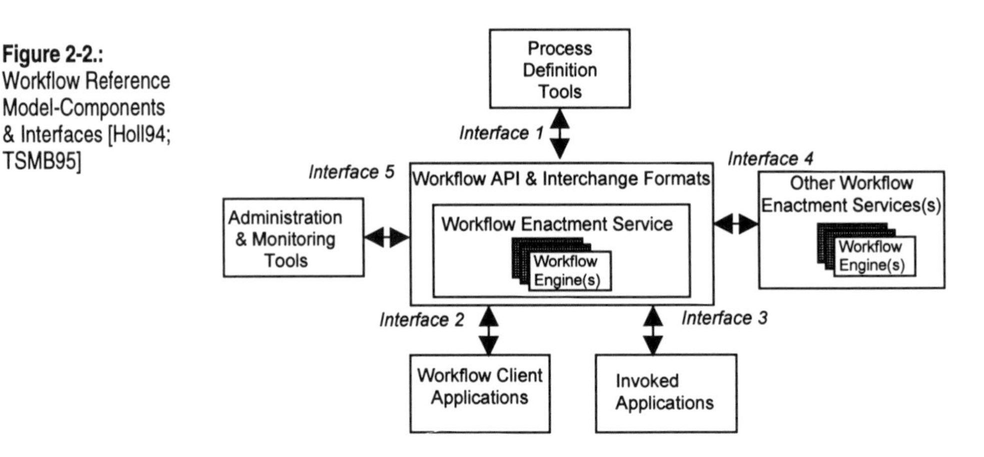

Figure 2-2 illustrates the major components and interfaces within the workflow architecture, these will be described as follows.

Workflow API and Interchange Formats

Workflow API and Interchange Formats – "The interface around the workflow enactment service is designated WAPI - Workflow APIs and interchange formats, which may be considered as a set of constructs by which the services of the workflow

[7] FlowMark is IBM's multi-platform client/server workflow management product for designing, refining, documenting, and controlling large scale, enterprise-wide business processes. For further information see [@IBM]

[8] Xerox's InConcert is a fully featured workflow management software that enables users to model and co-ordinate all components - people, procedures and documents - of a work process. InConcert' s open workflow architecture is designed specifically for flexibility and integration with other products. For detailed information see [@InConcert].

[9] SAP Business Workflow will be described in more detail in section 3.4

system may be accessed and which regulates the interactions between the workflow control software and other system components" [Holl94].

Workflow Enactment Service

Workflow Enactment Service – "The workflow enactment service provides the run-time environment in which process instantiation and activation occurs, utilising one or more workflow management engines, responsible for interpreting and activating part, or all of the process definition and interacting with the external resources necessary to process the various activities" [Holl94].

Workflow Engine

Workflow Engine – "A software service or 'engine' that provides the runtime execution environment for a workflow instance" [Holl94].

Process Definition Tools

Interface 1: Process Definition Tools – Defines a standard interface between the process definition tool and the workflow engine [PW97a], i. e. definitions for workflow icons and tools for the development of a common workflow [PW96a].

Workflow Client Application

Interface 2: Workflow Client Application – Defines standards for the workflow engine to maintain work items which are presented by the workflow client to the user [PW97a]. In other words it is a standard method for presenting work to a secondary process or an end user [PW96a].

Invoked Applications

Interface 3: Invoked Applications – Provides a standard interface to allow the workflow engine to invoke different applications [PW97a], for example calling a variety of third-party tools such as e-mail and document imaging [PW96a].

Workflow Interoperability

Interface 4: Workflow Interoperability – Defines a variety of interoperability models and standards relevant for each of them [PW97a]. It is a standard which allows diverse workflow products to share and interact with each other [PW96a].

Administration and Monitoring Tools

Interface 5: Administration and Monitoring Tools – This interface defines the administrating and monitoring functionalities which allow the passing of status information between workflow systems [PW97a; Holl94].

Please note that this concept could be described in more detail but that this is not the intention of this book. This chapter wants to set a common basis for further discussion on this topic.

2.4 CSCW, Groupware, Workgroup Computing and Workflow Management

At first glance, the differences between CSCW, Groupware, Workgroup Computing and Workflow Management are not immediately obvious.

A reason for this is the fact that their definitions are overlapping, which is demonstrated in the figure below.

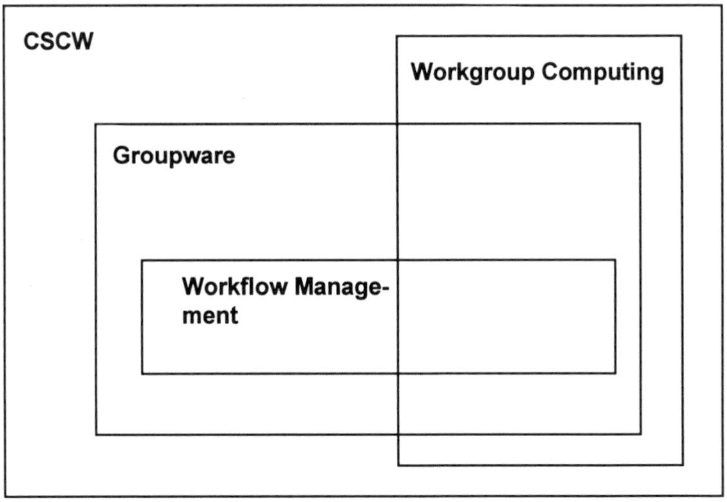

Figure 2-3: Overlapping of the terms CSCW, Groupware, Workgroup Computing and Workflow Management [Ste96]

In fact, CSCW, groupware, workgroup computing and workflow management systems have begun to converge into powerful programs which incorporate the functionality of all four terminologies.

In the chapters to come, we will refer to these four technical terms when we will discuss the product **Lotus Notes 4.5** and especially to its interface for invoking third party applications, like the **SAP R/3** system.

2.5 Lotus Notes 4.5

What is Lotus Notes?

According to the Lotus Development Corporation[10] Notes is built around a Client/Server architecture, where users (the Clients) communicate over a Local Area Network (LAN) or Wide Area Network (WAN) with document databases - or object stores - that are located on single or multiple Lotus Notes Servers.

Lotus Notes combines

- **Graphical User Interface (GUI),**
- **Application Development Environment,**
- **Document Database**[11]**,**
- **Sophisticated Messaging System** (e-mail),
- **Server Security Functionality,**
- **Customisable Templates for Common Business Applications** and provides
- **Support for Multiple Platforms** (e.g. Windows 95/NT, OS/2, AIX) and

gives the user the ability to create custom applications in order to improve the quality of handling business processes in areas like product development, customer service, sales and account management.

The Application Development Environment within Lotus Notes includes an integrated development environment, providing tools to developers of varying expertise.

The development environment of Lotus Notes includes several layers, which are the **Basic development facilities** like forms, views and templates.

Advanced programming facilities

Advanced programming facilities are formulas[12], @functions[13], macros and the Notes programming language Lotus Script with

[10] For more information, see [@Lotus]

[11] *Data* is the symbolic representation of states of affairs in the world, whereas a *database* is a collection of related *data items* mirrored as an abstraction of the reality. Each data item has a value, which comprise the state of the database at any time [Date95, DHLW96].

[12] Notes formulas are expressions that have program-like attributes. For example, the user can assign values to variables and use limited control logic. The formula language interface to Notes is provided through calls to @Functions [IBM96].

LotusScript Extension (LSX)

its add-on *LotusScript-Extensions* (LSXs[14]). In addition, it is possible to develop Notes applications by programming, using External Development Tools such as Lotus Notes *Application Programming Interface* (API) for the programming languages C, C++ etc.

Connectivity Extensions

Connectivity Extensions of the Notes development environment and its ability to contact external database systems, desktop applications, the Notes specific development tool (see Figure 2-4) and the Notes API.

Figure 2-4:
The Notes Integrated Development Environment (IDE)

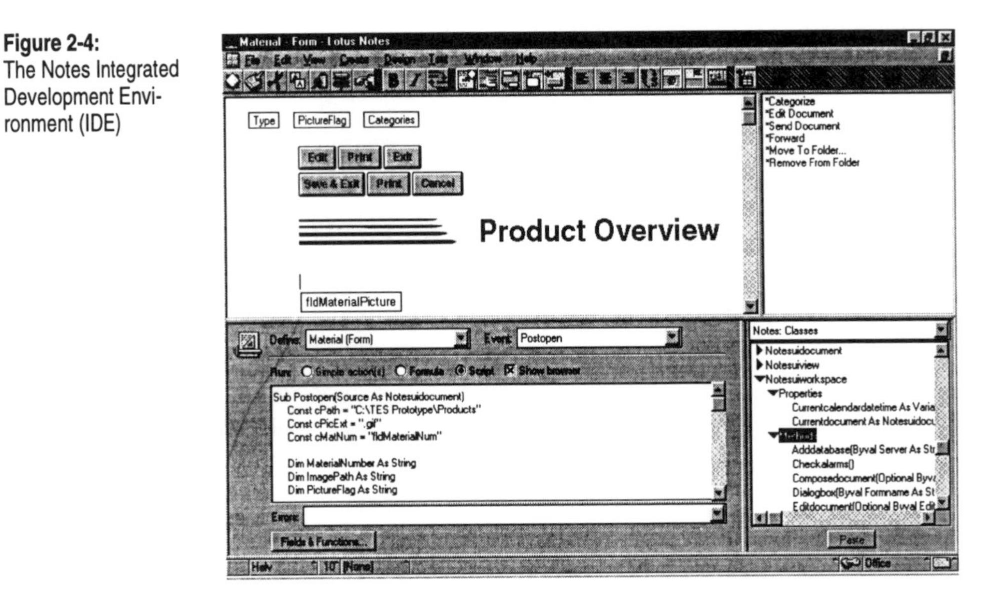

Integrated Development Environment

Lotus Notes is shipped with many types of design elements which are used to create a range of application types. The Notes *Integrated Development Environment* (IDE) is the single interface to all of the Notes application design elements. The main advantage of IDE is that the user only needs to learn how to use it once to access all Notes design features.

The IDE displays a three-pane work window, which is shown in Figure 2-4. The design pane in the top left half shows the form or view that is being created. The smaller action pane on the

---

13 @Functions are used to perform specific tasks within Notes. For example, @Created displays the create date of a document [IBM96].

14 For more information about LSX see chapter 6

right lists available actions. The programmer pane below is where the developer creates the code.

A key strength of the Notes application development environment is that the application design elements like forms, views, templates, macros, formulas, functions and procedures are stored as Notes documents. Thus replication services automatically manage application deployment.

Notes stores information in databases. A database is a file that contains *documents*, *forms* and *views* (see Figure 2-5). The term 'document' can often be misleading. Within Lotus Notes and its rich text field capability, a document can contain just about any 'electronic-object', which is why a Lotus Notes document is also known as a sophisticated storage container. The 'electronic-objects', which could be text, numerical data, tables, graphics, images, spreadsheets, OLE*, video files or a voice message that can be imported and embedded into a Lotus Notes document. Under Windows and OS/2 the user is able to link to or embed data from other Windows and OS/2 programs using DDE*.

In other words, the capability of attaching any file in any format to a Notes document for the distribution to others is another big advantage of Lotus Notes [Rich97]. All these files can be viewed straight from the document, detached or launched directly from the Notes application.

Form

A *form* is a *template*[15], which is designed and stored in the database, that can be used to create documents. The Notes forms are the building blocks of Notes databases, which are used by the user to enter information. The entered information is then stored as a document in the database.

View

A *view* is a list of documents sorted or categorised in different ways to support the user in finding information.

Security with Access Control List

Security within Notes is handled in a variety of ways. Users are granted or denied access to Notes servers through the certificates stored in their User ID-files. Each Notes database contains an *Access Control List* (ACL) detailing who can open the database and what they can do to its information. In addition, Notes in-

* For a detailed description see section 4.1.2

15 A template is a model of a document (which could be anything from a simple memo to a complex prospectus or contract). The template can contain boilerplate text, graphics, and formatting that all documents of that category commonly use.

formation can be encrypted so that only specified users can decrypt it [Rich97].

Figure 2-5: Notes Database Components

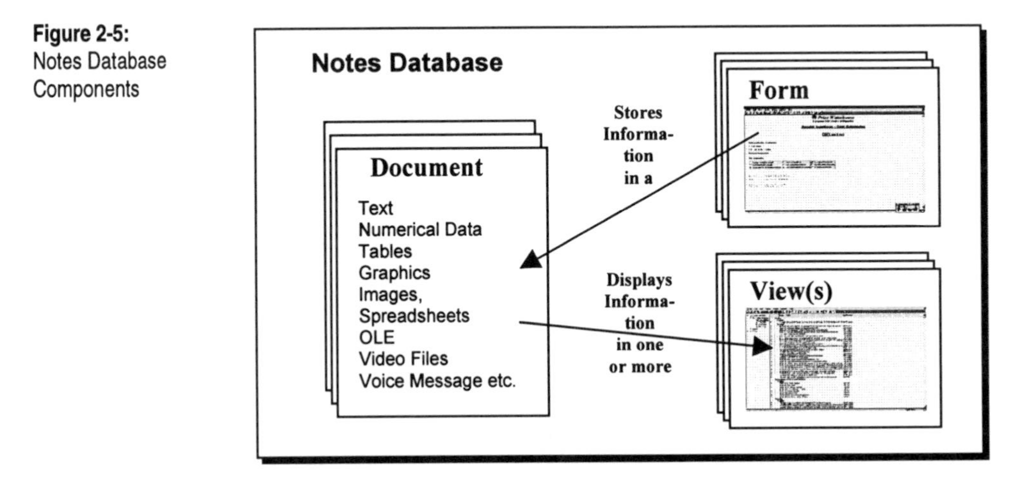

Synchronised Replication

Lotus Notes relies on *Synchronised Replication* to ensure that data on multiple Notes servers is consistent. Lotus Notes has gone some way for its release 4 towards adding 'web awareness' into its product.

With Lotus Notes' new capability to work with information on the Internet through its family product Lotus domino, it now combines the rich text and security capabilities of its package and the vast information of the World Wide Web – providing the best of both worlds [Rich97]. This will be discussed in more detail in the following section.

2.6 Benefits of Lotus Notes and Lotus Domino

The unique database structure of Lotus Notes is able to store complex, relatively unstructured information (which is how most information is typically retrieved) and then make this information available for certified groups and users on the company own network.

It does not matter if the single user is directly connected to the network or if they are dialling in from a remote location.

Lotus Notes considers also changes which are made by a mobile user who works with a Notes document in a so-called detached mode (not connected to a Notes network) and is able to make changes to the document on a record-level basis. This method reduces the amount of network traffic and ensures the consistency of each Notes database.

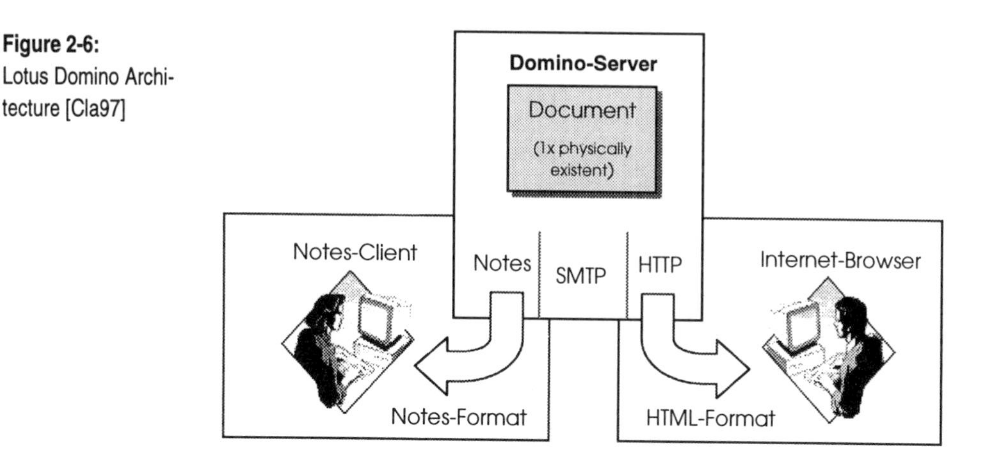

Figure 2-6: Lotus Domino Architecture [Cla97]

Lotus Notes helps to eliminate much of the redundant paperwork and steps in the company's business processes by moving the documents into an electronically organised workflow within the company.

By taking advantage of WWW-Standards and incorporating them into the Lotus Notes architecture with Lotus domino (see Figure 2-6), companies are able to leverage a single architecture to build client/server applications for internal use and to reach customers, business partners and suppliers.

This allows Web developers to use the Lotus Notes mature groupware functionality to reach non-Notes clients via the Internet, providing greater choice and flexibility [Rich97].

According to Lotus Development Corporation, "Domino is the server technology which transforms Lotus Notes into an interactive Web application server allowing any web client to participate in Notes applications securely. Bridging the open networking environment of Internet standards and protocols with the powerful application development facilities of Notes, Domino provides businesses and organisations with the ability to rapidly develop a broad range of business applications for the Internet and Intranet" [@Lotus].

3 Corporate Applications

Unlike CSCW, groupware, workgroup computing and workflow management, corporate applications cannot be discussed in a theoretical way as it was done in previous chapters. This is because they are in most cases 'grown' systems, which depend on the development methods of the various software companies and therefore will be described from a different perspective, i.e. their historical development.

3.1 The historical development

Originally many companies relied on internal application development to create a business application on which their corporational business was based. Their characteristics could be described as follows [PW96b]:

- They ran on mainframe computers.
- They were written in second- and third-generation languages.
- They automated functional processes, which have to be well understood.

"According to the book, *IS Organizational Transition*, over the last 35 years corporations world-wide have invested and estimated $2.7 trillion in these corporate applications" [PW96b].

Most often building custom applications on one's own is very time-consuming and therefore expensive and also very risky. Therefore most corporations prefer standard application packages which also include graphical application development tools for a rapid adaptation to changing business needs. Unless there are significant competitive advantages in inhouse-developments, these so-called off-the-shelf products are usually preferred [PW96b; PW97b; MW96].

As a result, the packages offered have to be designed as comprehensive business solutions to support a high degree of integration e.g. for multi-site and multi-currency operations, so that they can be adapted for any kind of company [PW97b; MW96].

In order to meet these requirements, there is now a trend to split the monolithic structure of these integrated applications up into several component based corporate applications (often called modules) which can be customised individually for each company [PW97b].

Looking towards the future the market for less-expensive applications will change the way in which corporate software is developed. New industry standards which define how systems are interoperating with third-party products and also how easily they can be configured and installed are the driving forces of a new kind of corporate software development [PW97b; MW96].

At present the major players in the software market such as SAP AG, Oracle Corp., Software 2000 Inc., PeopleSoft Inc. and Baan Co. "are developing standard-based, object-oriented software that will eventually make the jobs of business users and IT personnel easier" [PW97b].

3.2 Some facts about SAP

Founded in 1972, SAP AG[16] in Walldorf, Germany, is the leading global provider of client/server business application solutions. Today, more than 9,500 companies in over 100 countries have selected SAP client/server and mainframe business applications to manage comprehensive financial, manufacturing, sales and distribution, and human resources functions essential to their operations.

SAP AG has two major products on the market, the R/2 applications for the mainframe environment and the R/3 applications for open client/server systems. Both with the R/2- and the R/3-system, customers can decide to install the core system and one or more of the functional components, or purchase the software as a complete package.

The power of SAP software lies in real-time integration, linking a company' s business processes and applications, and supporting immediate responses to change throughout the whole organisation - on a departmental, divisional or global company level.

The real-time integration of R/3 is achieved through one central database system, which documents the entire R/3 system.

With more than 18,000 R/3 installations in over 100 countries [SAP98] R/3 is accepted as the standard in several industries, such as oil, chemicals, consumer-packaged goods, and high tech/electronics.

[16] SAP stands for: *Systems, Applications and Products in Data Processing*

SAP, Oracle and PeopleSoft are the three major players within the client/server corporate applications market. According to a survey that was conducted by the International Data Corporation[17], SAP held more than 29 percent of the market share in 1996 and therefore had the lead in the world-wide client/server market.

Looking back over SAP's 25-year history, SAP has evolved from a small software company to the world's market leader in the business application software area. Today, SAP is the largest supplier[18] of business application software in the world and the world's fourth largest independent software supplier, overall.

As we go along, the R/3 system and its architecture will be introduced in more detail.

3.3 SAP R/3 Functionality

The SAP R/3 client/server enterprise-computing package is a comprehensive business solution that includes powerful integrated tools for managing and using business data. More than 70 software modules are part of the R/3 family, including financial accounting, sales and distribution, human resources, and computer-integrated manufacturing (see Figure 3-1). R/3' s solid integration of its many modules and functions is its greatest strength.

R/3 has a two- or three-tier client/server architecture[19] compatible with most major operating systems and database servers from Oracle Corp., Informix Software Inc., Software AG and many others. The three-tier structure provides for efficient transaction processing and increased scalability. However, companies deciding to use R/3 must allow for a substantial commitment of time and resources to reengineer their business processes to be consistent with the R/3 system.

With the R/3 system, SAP is setting new standards for business information processing. The SAP family of products has the ability to model a wide range of business processes.

---

[17] International Data Corporation (IDC) is a leading provider of information technology data, analysis and consulting; with research centres in over 40 countries and more than 300 research analysts world-wide. Its goal is to provide a global perspective on IT market and technology trends. For further information see [@IDC]

[18] Compare information with SAP's Annual Report 1996

[19] For more about the Client/Server architecture of R/3 see section 3.3.1

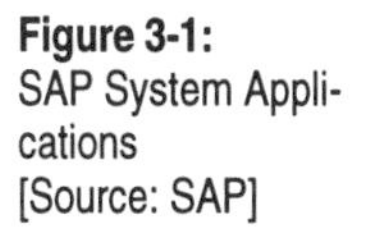

Figure 3-1:
SAP System Applications
[Source: SAP]

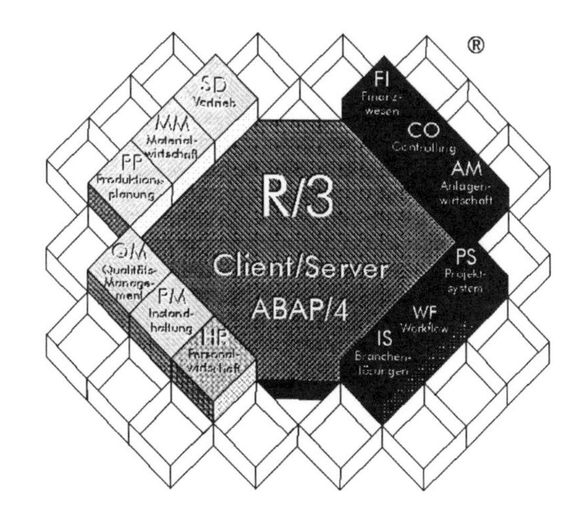

The products are designed for a international companies, with support for multiple currencies and local practices as well as support for distributed applications.

The R/3 system offers a complete infrastructure for corporate information processing. It includes mature standard business applications and tools for introducing the system and controlling and monitoring it during operation. The standard applications are designed using the ABAP/4 (Advanced Business Application Programming) Development Workbench, which is integrated in the R/3 System and is also available to customers for developing their own solutions and extending the capabilities of existing applications.

3.3.1 SAP Client/Server Architecture

The R/3 architecture is based on a software-oriented, multi-tier, client/server principle. Although SAP can be implemented on a centralised (one-tier) configuration, it usually utilises a highly distributed architecture, depending on customer requirements [BG96].

Two-tier R/3 configurations are implemented with special presentation clients that are responsible for the GUIs, generating what users actually see on-screen. For example, many R/3 users have PCs as presentation clients.

In a three-tier R/3 configuration, separate computers are used for presentation, running applications, and the underlying database.

Depending on customer requirements, users can have the option of connecting to one or multiple application servers. The application servers allow the processing power to be distributed over multiple computers [PW96b; PW97b].

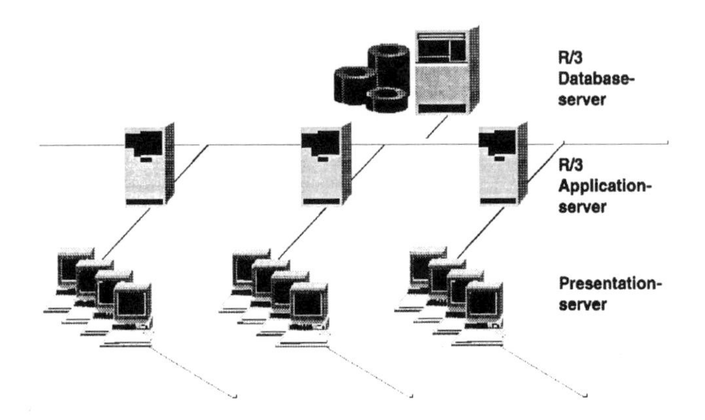

Figure 3-2:
The common three-tier model has established itself on the market
[Source: SAP]

Client front ends, which collect data for transactions and present results to users, run on a variety of operating systems, including UNIX/Motif, Microsoft Corporation's Windows, and IBM' s OS/2. The clients send transactions to application servers, which implement the business functions. In most cases, the database resides on a UNIX server, but SAP has announced plans to utilise the AS/400, MVS, and PowerPC platforms as servers.

The R/3 uses *Remote Function Calls* (RFCs)[20] for communication between clients and application servers and among application servers themselves. R/3 components communicate using major networking protocols such as TCP/IP, IBM' s SNA, and Novell Incorporation' s IPX/SPX [BG96].

The R/3 can be configured with multiple application servers. If one application server is unavailable, an alternate server is then selected automatically. Load balancing is achieved by assigning a client to the least-busy application server. The servers provide consistent caching of data related to a client request to reduce the load on the database server and provide improved response time for the client.

[20] This will be covered in much more detail in the BAPI chapter

3.3.2 SAP Development Environment

SAP has built a complete development environment that assists the maintenance and development of SAP R/3 applications. The ABAP/4 Development Workbench includes development tools, such as Screen Painter, Menu Painter, Editor, and Interactive Debugger; tools for performance measurements; and Computer-Aided Test Tools (CATT). It also includes the ABAP/4 Repository, the ABAP/4 Repository Information System, the Active ABAP/4 Dictionary, the Enterprise Data Model, and the ABAP/4 Development Organiser.

Customers have the flexibility of either developing their own customised applications or enhancing SAP-supplied application logic using SAP' s ABAP/4 fourth-generation language (4GL). All applications developed with the ABAP/4 Development Workbench are based on the SAP application development environment and are platform- and database-independent.

The ABAP/4 Development Workbench offers access to SAP development tools that cover all phases of the overall software development life cycle. Tools are available for writing business applications in ABAP/4, for accessing databases, for network communications, and for implementing GUIs.

3.4 Other SAP Products

This part will give you a short overview about additional products and functionality, delivered with a R/3 system to give a broader perspective.

SAP Business Workflow

SAP Workflow: The SAP Business Workflow concept is designed to accommodate the business processes of different functions of an organisation. Conceptually, it is based on the SAP R/3 Reference Model. SAP Workflow Management components provide software solutions for the operational level. The SAP Business Workflow concept provides support for transaction processing across all applications. It offers flexible control of business processes across transactions; worklist management for active, convenient administration of work-backlogs; event management for flexible definition of responses to events; deadline management for automatic starting of appropriate routines when required results are not obtained or when expected events fail to occur; linking of processes and objects of a transaction; and modelling, simulation, optimisation and monitoring of business processes.

A company's competitiveness mainly depends on its ability to process business activities effectively. SAP Business Workflow supports a company, which uses the R/3 system to achieve this goal.

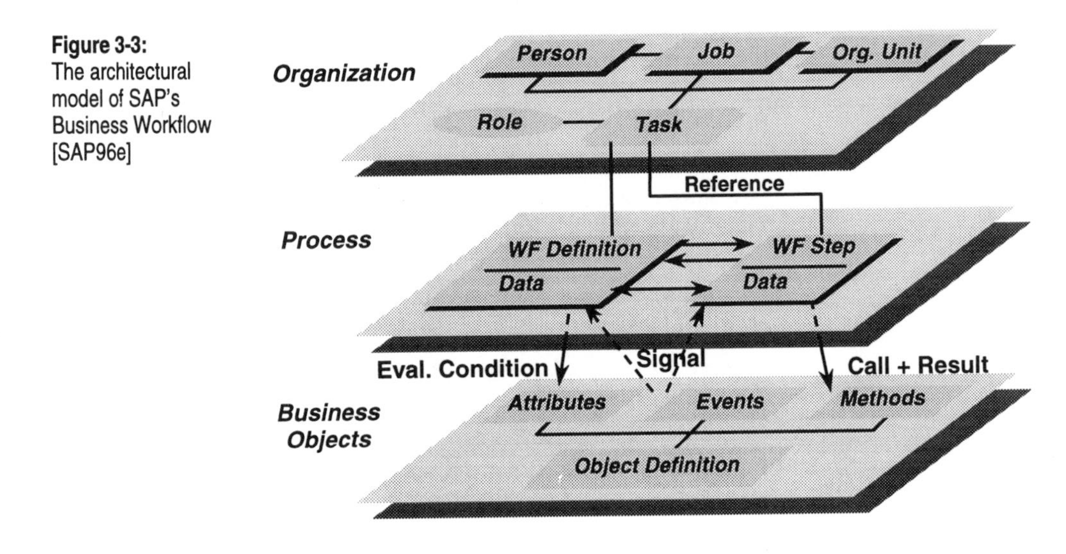

Figure 3-3: The architectural model of SAP's Business Workflow [SAP96e]

As shown in figure 3-3, SAP Business Workflow is a model-driven implementation approach. It is supported by SAP's Business Engineering Workbench (BEW) which comprises a set of tools and technologies to perform continuous business process engineering. The main components of the BEW are implemented models, which describe the various business processes and business objects of a company [SAP96e].

It also lets the user query the status of a business process at any time. Integrated organisation management makes the distribution of tasks to organisational units clear. These assignments of work allows substitution and successor situations to be handled in a dynamic and flexible way. Furthermore, a company can use this tool individually to implement additional processes across applications.

As a result, SAP Business Workflow supports business processes to be quickly and easily adapted to constantly changing market conditions. The product uses SAP Business Objects, which are stored in the R/3 repository. This object-oriented approach also allows SAP Business Workflow to link to external systems [Stro97].

The product permits fast and clear processing of business activities, including the possibility to model them from an Organisational, Process and Business Object level (see figure 3-3).

We will not go into SAP's Business Workflow in more detail, on an organisational and process level but we will look at SAP's Business Objects more closely in the chapters to come. For more information on SAP Business Workflow see [Stro97].

Application Link Enabling

Application Link Enabling (ALE)[21] – This product enables networks comprised of multiple SAP systems to be implemented. ALE uses asynchronous links to integrate business processes, avoiding rigid ties to a central database. For example, a company could run multiple SAP database servers and ALE would coordinate the flow of information based on predefined business rules.

Integration Support

Integration Support – The R/3 system is designed to be open. Development activities are based on internationally recognised standards. Among others, the R/3 system utilises the following open interfaces: TCP/IP, RFC, CPI-C, SQL, ODBC, OLE/DDE, X.400/X.500, MAPI, and EDI. For more detailed information about the interfaces of the R/3 system, see chapter 4.

[21] ALE will be described in much more detail in chapter 9.

4 SAP R/3 Interfaces

The following table shows the main interfaces which are provided by the SAP R/3 system. This is only a brief summary of the available R/3 interfaces, for a more detailed description see [Muth97; BG96] and the R/3 documentation. The GSS-API / SNC and the WAPI interface, have been added to Muthig's model, since their introduction in the R/3 Release 3.1G.

4.1 R/3 Interface Architecture

Muthig divided the different R/3 interfaces, based on Buck-Emden's model [BG96], up into *Object-Oriented*, *High-Level*, *Medium-Level* and *Low-Level* Interfaces, as shown in the following figure.

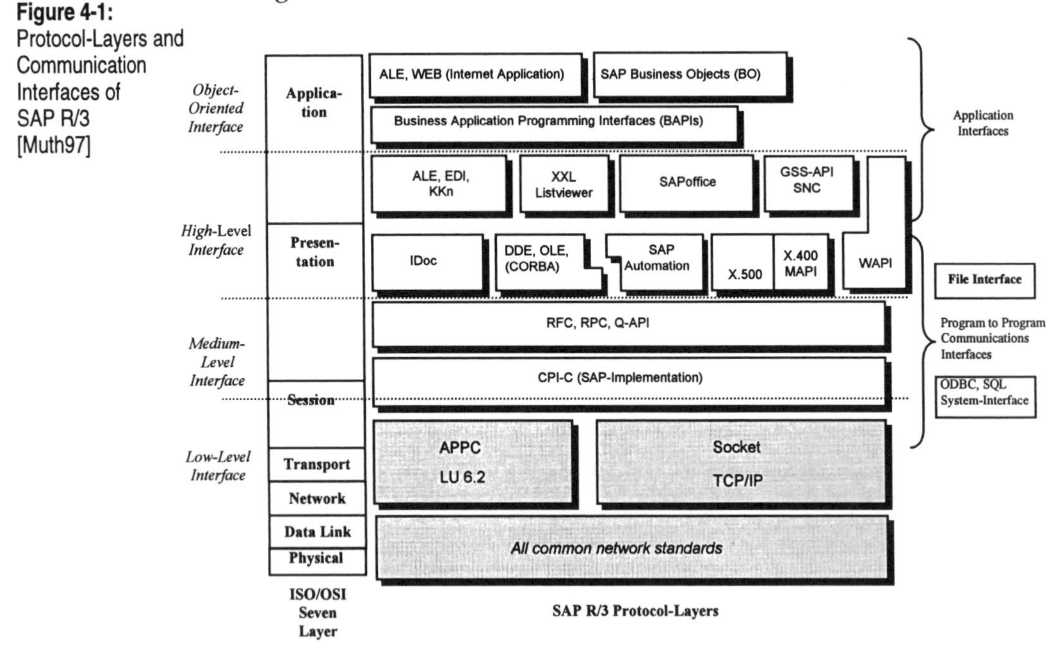

Figure 4-1: Protocol-Layers and Communication Interfaces of SAP R/3 [Muth97]

Note that the APPC and LU6.2 interface protocols are usually used in a mainframe environment, which is the common basis for the R/3-predecessor R/2. These interfaces were taken from the R/2 into R/3 to enable R/2-R/3 connections.

The Sockets and TCP/IP section form a common network protocol, which is used in a Client/Server environment typical of a R/3 installation. These grey-marked interfaces will not be described as this chapter will concentrate on the unmarked interfaces, which are the medium, high and object-oriented interfaces of R/3.

The focus in this chapter lies on a common R/3 implementation that would use the unmarked interfaces, described in the following tables.

4.1.1 Medium-Level Interfaces

CPI-C

CPI-C Common Programming Interface-Communication CPI (Common Programming Interface) is an IBM standard, which defines a standardised interface for program development. CPI provides languages, commands and calls for developing applications within a SAA[22] environment. CPI-C is the part of CPI, which provides definitions for inter-program communication. These definitions can be divided into four areas: Session setup, Session control, Communication and End of session

RFC

Remote Function Calls[23] enables the program to call and process predefined procedures (functions) in a remote SAP System. RFCs are managing the communication control, parameter transfer and error handling. RFCs are using TCP/IP in Client/Server- or SNA LU6.2 in mainframe- environments. The communication between two systems can be done synchronously (e.g. telephone call) or asynchronously (e.g. letter).

RPC

Remote Procedure Calls (RPC) allow program modules (procedures) running on other computers to be called.

Q-API

Queue Application Programming Interface is the interface which is commonly used by the buffered and asynchronous data transfer between the SAP systems R/2 and R/3. External systems can transfer their data to a buffer where the data will be read by a transmission program, based on RFCs which then transfers the data to another SAP system.

[22] Systems Application Architecture (SAA) is IBM' s family of standard interfaces that enables a software developer to write software independently of hardware and operating system [Eng93].

[23] For a much more detailed description see Chapter 8

4.1.2 High Level Interfaces

SAPoffice

SAPoffice is the e-mail system and folder structure in the R/3 System. The user can use the mail system to send documents internally (to other R/3 users in the same system) or externally (to users in other systems). SAPoffice enables the user to store documents in folders, which can be private or shared.

SAP Automation

SAP Automation is the term that describes a family of APIs currently being developed for the SAPGUI communications protocol. SAP Automation enforces programmers of complementary applications to directly access the flow of data for every SAP screen, from the application server to the enduser-system, and implement alternative user interfaces for telephony services, kiosk systems, WWW, etc. [SAP96b]

X.500

X.500 is the set of ITU-T[24] standards covering electronic directory services such as white pages and whois [IBM93].

X.400

X.400 The set of ITU-T communications standards covering electronic mail services provided by data networks. It is widely used in Europe and Canada [IBM93].

MAPI

Messaging Application Program Interface is Microsoft' s system interface for sending electronic mail across a local area network [Muth97].

XXL List-viewer

EXtended EXport of List – This XXL-tool is used to configure Microsoft Excel (or other spreadsheets like Lotus 1-2-3) for displaying and further processing of R/3 application data. The XXL List Viewer performs the following functions [SAP97b]:

- It provides specific functions (in addition to the standard functions of Excel) in a separate menu and function bar.
- It presents R/3 data in Excel, taking into account information delivered by R/3 on the structure of the data.
- Limits Excel functionality to guarantee the consistency of the data passed from R/3.

DDE

Dynamic Data Exchange is Microsoft's Windows hotlink protocol that allows application programs to communicate using a client-server model. Whenever the server (or "publisher") modifies part of a document which is being shared via DDE, one or more

[24] ITU-T, stands for International Telecommunications Union, Telecommunication Standardisation Sector [Eng93].

clients ("subscribers") are informed and include the modification in the copy of the data on which they are working [PW97a].

OLE, OLE 2.0

Object Linking and Embedding (also known as OLE Automation) is a distributed object system and protocol from Microsoft, which allows an editor to "farm out" part of a document to another editor and then reimport it. For instance, a desktop publishing system might send some text to a word processor or a picture to a bitmap editor using OLE. Microsoft is totally committed to OLE as the means by which Windows will evolve to a fully object-oriented operating system. OLE is a large and still growing specification designed to cover the entire range of behaviour of Windows objects.

The **OLE 2.0** specification defines, in a generic sense, what objects are and how they must behave in order to be called *Component Object Model* (COM) objects, how they are stored, how they communicate with one another and with their containers, plus embedding and linking, in-place activation, drag-and drop, and finally how objects can be controlled programmatically through OLE automation. It is a 16-/32-bit, platform-independent specification, equally applicable to Windows, Windows NT and successors, and to the Apple Macintosh. It even encompasses the potentially difficult issue of internationalisation (for example, how is the name of the colour property spelled correctly: color, colour, couleur, colore?) [PW97a].

OLE Automation

It is automation that brings OLE into contention with the SAP's R/3 System. OLE Automation means that any automation-aware client application can control OLE server objects. In terms of the R/3 system this means, that an OLE Automation server based on this RFC interface enables PC applications to use business objects similarly to other desktop objects.

CORBA

Common Object Request Broker Architecture is an Object Management Group specification that provides the standard interface definition between OMG-compliant objects.

ALE

Application Link Enabling[25] is the concept that supports the installation and operation of distributed applications. "It comprises an administratively controlled message exchange with consistent data holding in loosely linked SAP applications" [SAP96a]. In other words, ALE is able to link distributed SAP applications with one another, as well as with applications from

[25] ALE will be described in much more detail in chapter 9.

other vendors. Communicating between two applications is not done through a central database but through synchronous (request) and asynchronous (data exchange through IDoc) messaging.

IDoc

Intermediate Document The IDoc type indicates the SAP format that is to be used to interpret the data of a business transaction. An IDoc is a data-container which has the following structure, in other words consists of the following components:

1. A control record, which is identical for each IDoc type.
2. Several data records, where one data record consists of a fixed key part and a variable data part. The data part is interpreted using segments, which differ depending on the IDoc type selected.
3. Various status records, which are identical for each IDoc type. These records describe the statuses an IDoc has already passed through or the status an IDoc has achieved.

EDI

Electronic Data Interchange is used for the exchange of standardised document forms between computer systems for business use. EDI is part of electronic commerce. EDI is most often used between different companies or so-called "trading partners" [SAP97b].

KKn

Kommunikationskanäle (German for: Communication Channels) these communication channels are used for the communication of a R/3 system with technical subsystems (e.g. BDE[26], MDE[27], QDE[28] and NC). The communication is done with RFCs, IDocs and ALE-functions [Muth97]

*Note that the following three interfaces **GSS-API**, **SNC** and **WAPI** are new to the model which was introduced by Muthig. Because of that a more detailed description is given in the section to come.*

In supporting the security of distributed processing, the question of how to supply the highest possible level of security in distributed processing across networks of all kinds is of critical impor-

[26] The English translation of Betriebsdatenerfassung (BDE) is: 'factory data capture'

[27] Maschinendatenerfassung (MDE) stands for: 'machine data capture'

[28] Qualitätsdatenerfassung (QDE) is in English: 'quality data capture'

tance for many businesses. The R/3 Technology Infrastructure meets these security requirements by offering a variety of internal and external security mechanisms [SAP96d].

Figure 4-2: Integration of R/3 in network security products via the SNC Interface [SAP96h]

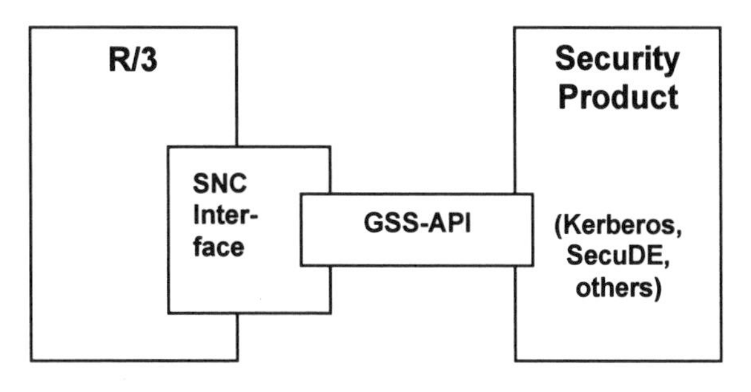

GSS-API and SNC

The first one is the standard ***Generic Security Services API*** (GSS-API) interface, which plays a key role in the implementation of a security mechanism for distributed processing. Because of that the GSS-API has been introduced as part of the R/3 ***System Secure Network Communications Interface*** (SNC) [SAP96d].

Kerberos

SecuDE

The R/3 System can be integrated with any network security product that itself supports the GSS-API; examples are *Kerberos* from MIT (Massachusetts Institute of Technology, USA), and *SecuDE* from the GMD (Gesellschaft für Mathematik und Datenverarbeitung, Germany). These systems allow external R/3 users to be authenticated and guarantee security of the communications data [SAP96d].

In addition to this, most R/3 users who operate their applications on public networks have installed firewalls. The R/3 System also offers its own sophisticated functions for controlling the access to data and functions, in order to prevent unauthorised access.

Secure Electronic Transaction

Of particular relevance to Internet applications is the *Secure Electronic Transactions* (SET) standard, which SAP plans to integrate into the R/3 System. SET was developed by a consortium of technology providers and credit card companies and is based on the *Private Communication Technology* (PCT) security package of Microsoft and the *Secure Socket Layer* (SSL) package of Netscape. These packages were designed to address the problems of client authentication, server authentication, confidentiality, connection reliability and the security of payment transactions [SAP97a].

WAPI

Workflow Application Programming Interfaces – These interfaces had been introduced to enable customers to link a R/3 system and its internal workflow capability with other systems, SAP has decided to publish its own APIs for workflow.

The SAP Business WAPIs include:

- Interfaces through events, which means that an external system can start workflows in the R/3 system through events. Events triggered in the R/3 System can cause a response in external systems (see figure 4-3) and also vice versa.
- Interfaces to generate and process work items.
- Interfaces to generate and execute workflows.
- Interfaces to use external clients inbox (for example, MAPI).

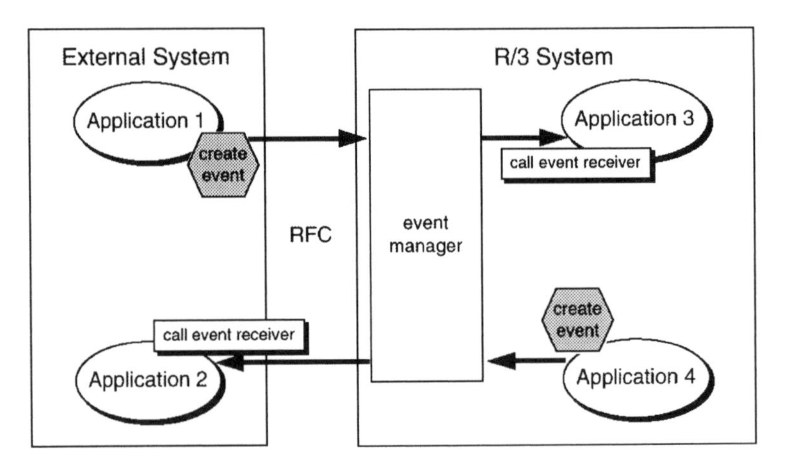

Figure 4-3: Communication through events [SAP96f]

Parallel to the introduction of the WAPIs, SAP introduced various interfaces (as we have seen in chapter 4.1) to connect the R/3 application with external systems.

At first it seems that we have discovered a controversy in the interfacing strategy of SAP, but it has a major advantage for the user. The user is able to specifically connect his/her R/3 system with external systems independently from the workflow system. In addition to that these interfaces are used by the workflow system.

Figure 4-3 visualises the possibility of establishing a communication between a R/3 and an external system through an application, which creates an event and establishes a connection via a

RFC call. We will take a closer look at the RFC technology in chapter 8.

4.1.3 Object-Oriented Interfaces

BO

The **Business Object (BO)** is a key R/3 System object. Data models are assigned to business objects. They describe the inner structure of the business object from the data view. A Business Object is a representative for a specific Business Entity like a customer order object.

BOR

The **Business Object Repository** (BOR[29]) is a part of the R/3 repository. The BOR contains all the meta-data of all interfaces, i.e. Business Objects [SAP96c] and BAPIs.

SAP Repository

The R/3 repository contains all data definitions, dynpros[30] ABAP/4-programs, reports, documentation, help files and the data-dictionary-data [Muth97].

ALE/WEB

ALE/WEB is the expansion of ALE through adding new distribution scenarios and *Internet Application Components* (IACs) to this interface [Muth97; SAP96f].

BAPIs

Business Application Programming Interfaces are the methods for accessing the aforementioned Business Objects, which will be described in their own Chapter 8.

[29] For more information see Chapter 7

[30] A dynpro (abbr. for: **dyn**amic **pro**gram) consists of a screen or mask with an associated processing logic including validation procedure. One dynpro controls exactly a single dialog step [SAP96e].

5 Lotus Notes and SAP R/3

Today, enterprises around the world face increasing competition in their markets, where the key to success is the ability to collect and broadcast information quickly throughout a global organisation.

The combination of Lotus Notes and SAP R/3 can add substantial value to an enterprise's ability to provide accurate information when and where it is needed. Notes and R/3 working together also makes it possible to extend information sharing to enterprise Intranets and the global Internet.

This chapter will try to describe the possibilities, capabilities and the tools, which are available today for the Lotus Notes and SAP R/3 integration.

5.1 Lotus Notes Database versus a RDBMS

"Lotus Notes is a document-oriented database and application environment that focuses on secure, distributed, collaborative group processes workflow, and unstructured information. RDBMSs and transaction systems are designed to help an enterprise in building high-volume transaction processing applications" [Lotus96b].

The key differences between Notes applications and DBMS applications are summarised in the following comparison.

Data Management

RDBMS – Captures, manages and shares an organisation's structured data.

Lotus Database – Captures, manages and shares an organisation's semistructured information, including rich data objects such as voice and graphics.

Update Capability

RDBMS – Implemented immediate access or updates to a data source through SQL[31].

Lotus Database -- Controls distributed operations with periodic updates via Notes field-level replication, "a powerful and cost-

[31] Structured Query Language (SQL) – A language that provides a user interface to relational database management systems, developed by IBM. SQL is the de facto standard for RDBMS as well as being an ISO and ANSI standard [Eng93].

effective means" [Lotus96b] of supporting distributed operations and occasionally connected users.

Concurrency Control

RDBMS – Control concurrent operations by using locking and isolation levels on database tables that ensure database integrity.

Lotus Database -- Concurrency control and occurring concurrency problems in a Notes environment are known by the term replication or save conflicts. Notes resolves that conflict by creating a so-called conflict document. If Notes cannot resolve the conflict automatically, one of the two document owners (normally the owner of the child document) has to decide which is the right document version.

The Underlying Network Infrastructure and the ability of Sharing Databases

RDBMS – The components used to build the RDBMS network infrastructure have to be determined. Sharing the data through shared database systems and applications, for example with suppliers, customers, and vendors may require the outside users to conform to a pre-determined set of network specifications. Web functionality is added through new tools to existing RDBMSs.

Lotus Database -- Notes databases and the network infrastructure, which make them available, are not restricted to a predetermined set of components. Through Notes and Web clients, Notes applications can be shared easily and securely with customers, suppliers, and vendors with diverse systems and environments, facilitating greatly simplified inter-enterprise application sharing. This is supported through the built-in Web-server functionality in Lotus domino. It allows internal network components to be changed "underneath" the Notes infrastructure to provide flexibility and management of the network separately from the Notes platform.

Database Design

RDBMS – Have rigorously be defined as on-storage physical and logical database schemas, requiring designers to translate business' terms into highly structured DBMS domains and entities, which is done through Entity-Relationship-Modelling technique.

Lotus Database -- Permits the designer to utilise flexible data modelling for applications. In other words there is no existing modelling technique for designing Notes databases. The Notes database design is done implicitly through the design of its forms and views, respectively documents.

Security

RDBMS' security features typically extend the features offered by the underlying operating system. Database security addresses the following areas; authentication, access control, authorisation and auditing [PW96c]. In other words modern RDBMSs ensure proper user authentication, guarantee the privacy and integrity of data, manage the assignment of database privileges, and monitor database operations across the enterprise computing environment.

Lotus Database -- Notes supports different security levels on server, database level, workstation level, form and view level, document level and field level [IBM96].

As probably already seen, a key distinction between Notes and a RDBMS is difficult because Notes does not provide capabilities usually associated with RDBMSs, such as the mentioned referential integrity and distributed transaction support.

However, Notes and RDBMSs are not mutually exclusive; in fact, they can exchange data and thereby extend each other, creating the possibility for very powerful applications. For instance, a manager might use Notes to create a monthly business report. Part of that report would pull in figures from a monthly expense database, based on a RDBMS, using an interchange tool. There are already some tools on the market and some of which are implemented in the two products, for a direct data-interchange between them. We will consider these tools in more detail in section 5.3 of this chapter.

5.2 The combination of Notes and R/3

Each of the two products has a unique design model and architecture that has different strengths when applied to their particular task.

The R/3 system, for instance, is designed to be the backbone of the companies business and to hold all core data of their day-to-day business and operations.

Lotus, as mentioned in chapter 3, has been a pioneer in developing software, which is able to empower people to communicate and collaborate more effectively and efficiently by providing a messaging, workflow, groupware and Internet backbone for a company.

As companies use their R/3 system, the need arises to make data of a R/3 system more accessible to enhance its business value.

"The power of Notes comes from its ability to support 'soft', unstructured information and allow users to access and share it as appropriate. Notes provides messaging and collaborative capabilities that support the unstructured, ad hoc data and workflow environment prevalent with today's knowledge workers" [Rich97].

The information that was discovered about Lotus Notes and a RDBMS leads to a direct comparison between Notes and the R/3 system, which will be shown in the following table.

	SAP R/3	Lotus Notes
Environment	Transaction Oriented	Documents and Intelligent Forms oriented
Data	Highly Structured	Unstructured Data, Rich Data Types
Database	Relational Database System, Centralised	Distributed, Replicated
Data Processing	Online Processing	Mobile, Remote and Networking
System Scope	Enterprise Scope	Workgroup towards Enterprise
Processes	Defined Business Processes	Defined business processes and ad-hoc processes can be supported.
WWW Capability	Internet Enabled Applications	Internet Server / Development Platform

5.3 Integration Technologies

There are different forms of integration technologies possible between Notes and R/3 that can be divided into the following categories [Lotus96a; IBM96]:

(a) End User Interfaces,

(b) Direct Database Access,

(c) Developer Tools and

(d) Programmable Interfaces.

5.3.1 End User Interfaces

End user interfaces are easy-to-use tools that give end users direct access to R/3 information with minimal to no setup or coding required.

The various R/3 interfaces have been discussed in chapter 4 and now the developed end user interfaces of Lotus for a R/3 connection, will be described as follows.

Lotus Connection for SAP R/3

Lotus Connection[32] for SAP R/3 – The Lotus Notes Connection for SAP R/3 allows end users to analyse, enhance and distribute R/3 report data through the various office products of Lotus, Notes and Domino. The data exchange is done by R/3's standard XXL data export facility.

Mail Support

Mail Support – Lotus Notes is providing a group of so-called *Message Transfer Agents* (MTA) that supports industry standard messaging protocols like X.400 and SMTP[33]. This allows messages to be routed between Notes and R/3. In addition, Notes is supporting MAPI as a server and allows MAPI clients to access Notes mailboxes. In other words, as mentioned before in this book, SAP is supporting MAPIs as well, so that a connection between Notes and the R/3 can be established, where a R/3 user can use Notes as the messaging infrastructure.

5.3.2 Direct Database Access

Notes is providing two tools to integrate relational database management systems (RDBMSs) with a Lotus Notes application. Since the R/3 system is storing its data in a RDBMS, Notes is able to access and mirror tables in the following two ways.

Lotus NotesPump

Lotus NotesPump – Lotus developed NotesPump as an enterprise server engine with the main task to transfer data between Lotus Notes and RDBMS data servers. It includes the following built-in features like SNMP[34] management, server load balancing,

[32] This program is available free of charge on the SAP R/3 CD.

[33] Simple Mail Transfer Protocol (SMTP) is a protocol, which is used to transfer electronic mail between computers, usually over Ethernet [Hunt93].

[34] Simple Network Management Protocol (SNMP) is the Internet standard protocol, which is developed to manage nodes on an IP network. SNMP is not limited to TCP/IP. It can be used to manage and monitor all sorts of equipment including computers, routers, wiring hubs, toasters and jukeboxes [Hull93].

activity fault recovery, scaleable server architecture and an automated control for storing backups.

LotusScript:DataObject (LS:DO) – The LotusScript:DataObject (LS:DO), supplied with Notes Release 4.5 is another available data integration method. LS:DO objects can access (read, write and query) any database for which an Open Data Base Connectivity (ODBC[35]) driver exists. Accessing a database consists of three steps: making the connection, specifying a query and getting values from the result set. LS:DO has three object classes that relate to these functions.

5.3.3 Developer Tools

Developer Tools give programmers powerful links to build applications that link R/3 to Notes and Domino. Lotus Notes provides developers with a cross-platform, integrated development environment that includes:

MQSeries

MQSeries is IBM's messaging middleware layer to link Lotus Notes and R/3 applications. MQSeries, used with SAP's ALE technology allows the exchange of encapsulated-data, transaction information and IDocs between these two systems.

LotusScript

LotusScript – Notes' LotusScript* is an embedded, object-oriented, BASIC-compatible structured programming language, which is similar to Microsoft's Visual Basic™. Furthermore, LotusScript is able to access external systems like R/3, which can be enhanced via LotusScript Extensions, or LSXs*, which encapsulate the function of the external system.

5.3.4 Programmable Interfaces

SAP R/3 and Lotus Notes are providing a number of different programmable interfaces to support the development of workflow, web and other collaborating applications. The programming interfaces are:

---

[35] ODBC stands for Open Database Connectivity and is a standard for accessing different database systems. "ODBC is based on the Call-Level Interface (CLI is a programming interface designed to support SQL access to databases from other application programs. CLI was originally created by a subcommittee of the SQL Access Group) and was defined by the SQL Access Group. Microsoft was one member of the group and was the first company to release a commercial product based on its work (under Microsoft Windows) but ODBC is not a Microsoft standard (as many people believe)" [DHLW96].

* LotusScript and its LotusScript Extensions will be described in further detail in chapter 6.

- **OLE Automation**
- **Remote Function Calls** – Notes can use its programming language LotusScript and also OLE Automation
- **SAP Automation Server** – The Automation Server allows R/3 transactions to be scripted from an external application like Lotus Notes. This interface is available through R/3's C-API interface as well as through OLE and can be launched by a LotusScript function.
- **ABAP/4** – Programmers can access the full range of Notes functionalities from within R/3, using the OLE Automation interface. This allows the ABAP programmer to open Notes databases, create documents, add and alter data, etc.
- **BAPI Interface** – With LotusScript, programmers are able to access R/3's Business Application Programming Interfaces (BAPIs). This SAP technology is a stable, object-oriented framework and enables the user to access Business Objects within the SAP system from remote sources such as Lotus Notes.

5.4 Possible Scenarios

There are many possible scenarios where the combined power of Notes and R/3 can enhance the productivity of an organisation. Some of the beneficial areas could be, for example:

- Business Web Sites,
- Remote Sales Force Automation,
- Access to HR-Information through Employees,
- Customer Service and Support,
- Distribution of Business-Reports and Data,
- Expense Reporting and Tracking and
- System Configurators.

The **Remote Sales Force Automation** scenario has been chosen to represent the potential of a Lotus Notes and SAP R/3 connection. This prototype scenario will be introduced in chapter 10.

6 LotusScript and LotusScript Extensions

"LotusScript is a BASIC compatible embedded scripting language with object-oriented extension. LotusScript first appeared in Lotus Improv, which was a Lotus' spreadsheet product, released in 1992 for Windows. Gradually, more Lotus products were shipped with LotusScript, including Lotus Forms and Lotus Notes ViP" [Rich97]. The actual LotusScript version that is used in Lotus Notes 4.5 is its Release 3.

6.1 Object-Oriented Terminology

LotusScript uses the term "class" to describe the "contract" defining the properties, methods and potentially the events related to an object. Notes uses an object-oriented, event-driven programming model. The interfacing between LotusScript and Lotus Notes can be done through a set of Notes object classes (see Figure 6-1). "Objects of each class generate and respond to different events. For example, a button object can respond to being clicked, or a field object can respond to the cursor being placed in the field" [Rich97].

LotusScript is capable of the basic object-oriented (OO) development features, which are Data **Encapsulation, Inheritance**[36], **Polymorphism** and creating **Classes** [Rich97; Boo94].

6.1.1 Encapsulation

"Encapsulation is most often achieved through information hiding, which is the process of hiding all the secrets of an object that do not contribute to its essential characteristic: typically, the structure of an object is hidden, as well as the implementation of its method" [Boo94].

In other words it is the ability to provide users with a well-defined interface to a set of functions in a way that hides their internal workings. In object-oriented programming, encapsulation is the technique of keeping data structures and the methods (procedures), which act on them, together.

[36] The OO-technique of inheritance and polymorphism is described in more detail and with an example in chapter 7.3.3

6.1.2 Inheritance

Inheritance is in object-oriented programming, the ability to derive new classes from existing classes. A derived class ("subclass") inherits the instance variables and methods of the base class ("superclass"), and may add new instance variables and methods. New methods may be defined with the same names as those in the base class, in which case they override the original one [Boo94]. For example, bytes might belong to the class of integers for which an 'add method' might be defined. The byte class would inherit the 'add method' from the integer class.

6.1.3 Classes

A class, in general, can be thought of as a template that can be used to create multiple instances of objects at run time.

Figure 6-1: Lotus Script Classes in Lotus Notes 4.5 [Source: @Lotus]

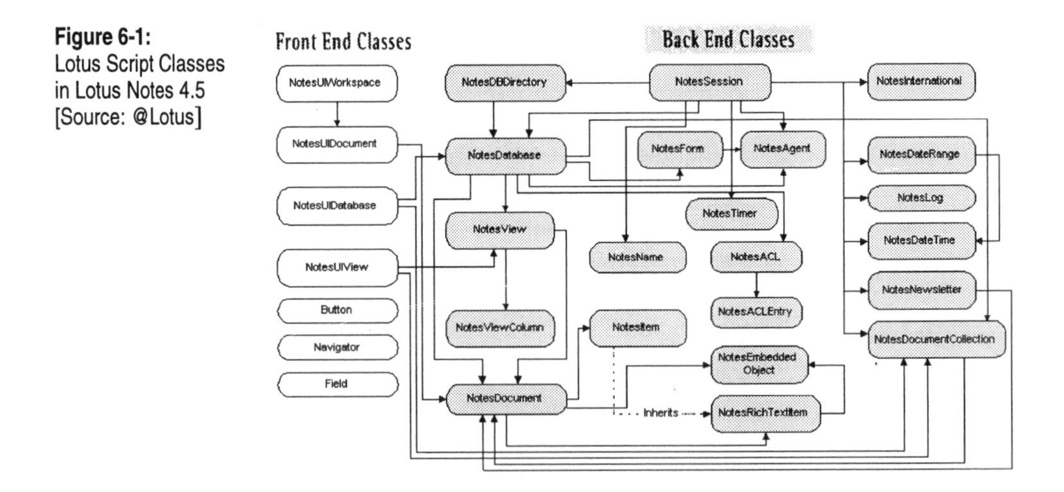

Object

Each object contains its own copy of data, which can be changed via properties provided externally to the user (or developer). The objects also contain code, also known as methods, which are used to manipulate or inspect this data at run time [Boo94].

Front-End and Back-End Classes

Notes has introduced, as shown in figure 6-1 two types of classes that enable the user to work with Notes objects. These are the front-end or user interface (UI) classes and the back-end classes.

The UI classes enable work with the databases, views and documents, displayed in the active Notes window. For instance, the user can work with the document that is currently on-screen using the NotesUIDocument class. The methods and properties of the UI classes support work with views and documents as a user would do it during a Notes session.

The second type of class, which is the back-end class, represents the constituent parts of Notes. This class includes some obvious elements such as databases, views, documents, fields and some more abstract concepts such as sessions and collections of documents.

Both types of classes are interlinked, which means for example a document displayed on-screen has two different representations – the UI document and the back-end document. When the user types data into the UI document, the corresponding back-end document is not updated until he/she saves the changes. Conversely, if an agent[37] updates the back-end part of a document, the UI document does not display the changes until it is refreshed.

The two types of classes, which were described in this chapter, are not the only classes available from LotusScript. Other types of classes can be added by a developer or by Lotus as a class collection within a *LotusScript Extension (LSX)*, explained in more detail in the following sections.

6.2 The LotusScript Extension for SAP R/3

The Lotus Script Extension (LSX) exposes new classes to the Lotus Script development environment. The user can use these additional classes in exactly the same way as he/she would use native Notes or SmartSuite product classes. The IDE also displays the new classes in the object browser and enables the user to look at run-time objects during a Lotus Script debugging session [Lotus97]. Examples for available LSX are classes to establish ODBC- or RFC-connections to external systems.

37 "Agents can work in the background to make things happen" [Rich97], in other words they are macros in Notes R4.x. Agents might automatically send mail, move documents and so on.

The 'LSX for SAP R/3' enables bi-directional access to R/3 data. In other words with the LSX the user is able to build applications which read and/or update data in a SAP R/3 system.

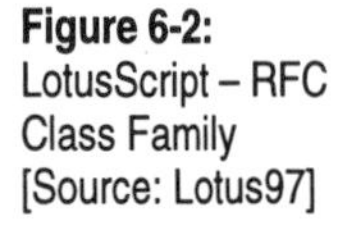

Figure 6-2: LotusScript – RFC Class Family [Source: Lotus97]

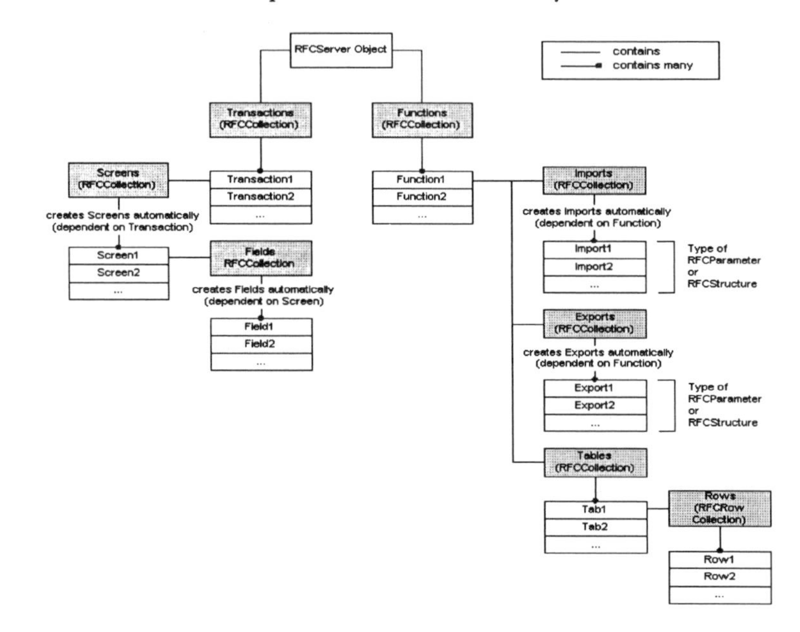

Lotus based this development on the RFC technology of SAP and introduced a separate RFC Class family, which encapsulates this technology into LotusScript.

Figure 6-2 shows the RFC Class family of the LSX for SAP R/3. This will be described in more detail, when we take a closer look at the coding of the prototype application.

The practical usage of the above mentioned classes will be described in more detail within the technical prototype description, in chapter 11 of this book.

7 Business Framework Architecture

In the mid 90s, SAP was heavily criticised by press and research companies because of R/3's monolithic structure, which was inherited from R/2, and because of its limited openness against other products. This became even more of an issue as SAP's major competitors like PeopleSoft and Baan announced in the meantime, that their products would be object-oriented. As a response SAP assigned an enormous research and development budget to redesign their R/3 system and to provide more flexibility for their customer base. Since the end of 1996 and the beginning of 1997 it seems that a big redesign and rewrite process is underway, to prepare the R/3 product for the future. The solution, which they came up with, is known as SAP's *Business Framework Architecture* (BFA).

SAP's definition of their BFA is: "The Business Framework is the result of many years' endeavour driven by customer requirements. Its foundation is R/3's original 3-tier client/server open architecture. The strategic development of interfacing and integration technology as well as the strong emphasis on easier-to-handle functional components led to the Business Framework" [SAP96c].

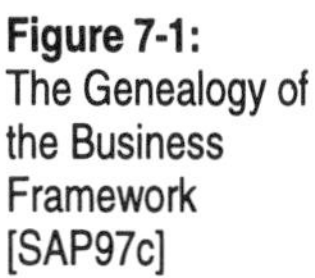

Figure 7-1: The Genealogy of the Business Framework [SAP97c]

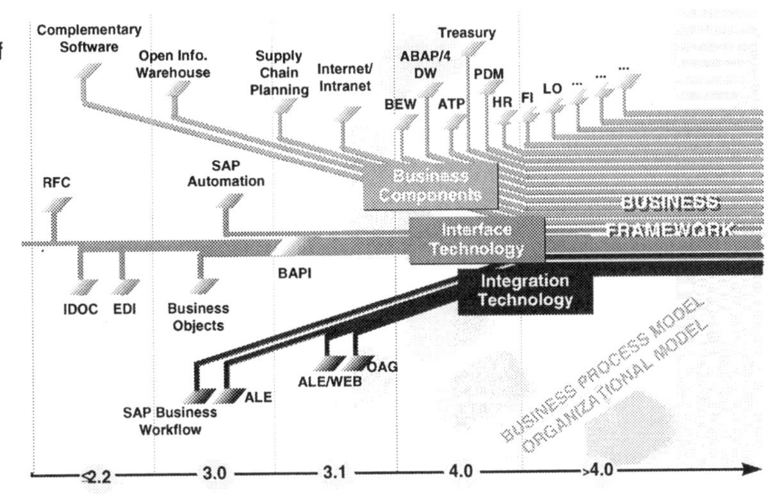

In other words the Business Framework is SAP's strategic product architecture on which R/3 is continuously developed and shipped.

The major aim of SAP's BFA is to facilitate and accelerate R/3 implementations. Figure 7-1 illustrates the genealogy and the components, which are the core of the BFA, whereas figure 7-3 gives a technical overview how the individual business components can be interfaced using the BFA's interfacing technology, which will be described in much more detail in the next chapters.

7.1 Business Components and BOR

The SAP *Business Components*, *Interface Technology* and the *Integration Technology* are essential parts of the Business Framework and the prerequisites for interoperability. This open, component-based architecture allows software components of SAP and third parties to interact with each other.

Business Components

The *Business Components*, like e.g. the HR, FI modules, encapsulate a specific functionality which can be accessed only using the interface of the Business Component's Business Objects (see Figure 7-3). In other words, a set of Business Objects, like 'Material', 'CustomerOrder' etc. is assembled within the R/3 system into a single business component, which is called *Logistics General*.

Figure 7-2: Structure of a Business Object in the Business Object Repository [SAP97,c]

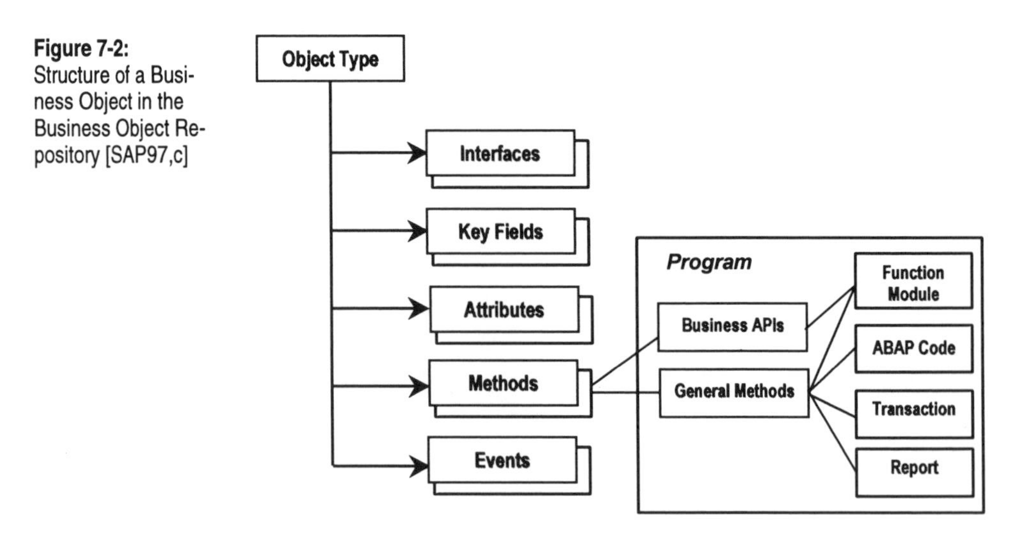

Business Object Repository

The *Business Object Repository* (BOR) is part of the R/3 Repository, which forms the 'documentation' of the R/3 system. The R/3 Repository contains a detailed description of the R/3 applications and is able to export information via an API to graphics software, modelling tools or *Business Process Reengineering* (BPR) tools. "The Repository is the central container for all of R/3's application information, including new developments, design and maintenance of applications and other components" [SAP96c].

All SAP Business Object types and their methods are identified and described in the R/3 BOR. The BOR contains all *meta-data* about the SAP Business Objects. In data processing, **meta-data** is definitional data that provides information about or documentation of other data, managed within an application or environment. For example, meta-data would document data about data elements or attributes (name, size, etc.), data about records or data structures (length, fields, etc) and data about data (where it is located, how it is associated, etc.). Meta data may include descriptive information about the context, quality and condition, or characteristics of the data [Boo94].

With the R/3 Release 3.0, SAP introduced its Business Objects and Business Workflow. Since then, "the BOR" and Business Objects "has been used primarily by the SAP Business Workflow" [SAP96c; SAP97d].

The BOR contains two types of objects [SAP97d]:

Business Object Types

Business Object Types – generally speaking represent SAP's Business Objects, which will be discussed in section 7.3 of this chapter. SAP structured and classified its R/3 Business Objects in different business application areas, such as logistics, sales and distribution, controlling, etc.

Technical Object Types

Technical Object Types – These are items such as text, work items, archived documents as well as development and modelling objects.

The BOR, 'plays' a major role in handling the access of Business Objects within the R/3 system and it serves, in this context the following essential purposes:

- "It identifies and describes the available SAP Business Object types and their BAPIs.
- It creates SAP Business Object instances. The runtime environment of the BOR receives requests to create runtime objects from client applications and creates the appropriate object instances. The client can then invoke the available BAPI methods on the runtime objects" [SAP97c].

7.2 The SAP Business Object

SAP has introduced object-oriented technology into the R/3 system by making R/3 processes and data available in the form of SAP Business Objects (BO). They cover a broad range of R/3 business data and processes that can be accessed through stable and standardised interface methods, which are called Business Application Programming Interfaces (BAPIs). SAP Business Objects and their BAPIs thus provide an object-oriented view of R/3 business functionality.

The whole concept of object technology and business object programming within the R/3 system is based on 'Real-World Business Objects'.

Figure 7-3: The Business Framework [Source: SAP]

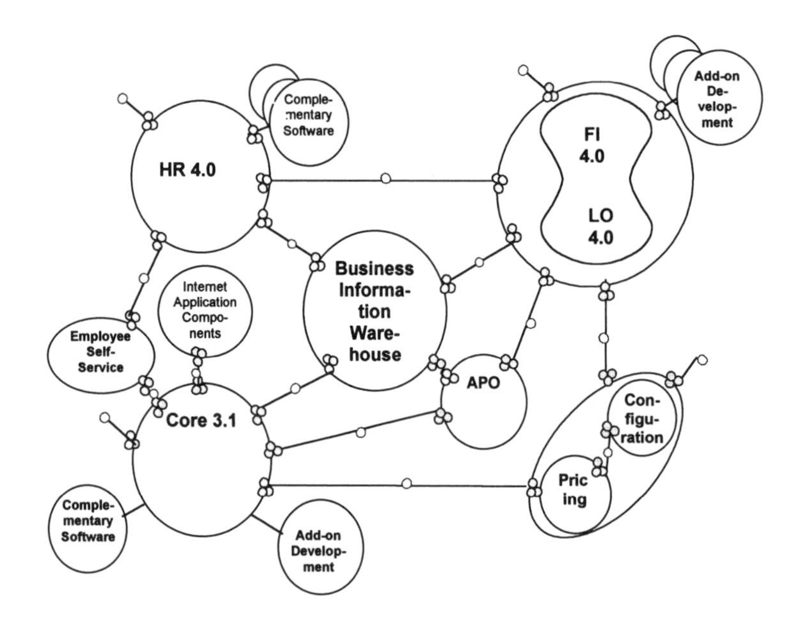

"The main emphasis of SAP's object strategy and vision is to show integration and interrelationships of objects on a business level, and not only on a technical level" [SAP96c]. Real world Business Objects like '**Material**' are modelled in the R/3 system and stored in the BOR. The characteristics of SAP Business Objects are that they implement standard Object-Oriented-Technology capabilities like *encapsulation*, *polymorphism* and *inheritance* [SAP96i]. We can visualise a Business Object, according to the object-oriented paradigm encapsulation, as a "black box" that encapsulates R/3 data and business processes and hides all details of its data structure and implementation to the user. To achieve this encapsulation, the SAP Business Objects are constructed as entities with multiple layers as can be seen in figure 7-4. SAP is dividing its Business Objects into four layers [SAP97c]:

- **Access Layer** – This layer defines the technologies that can be used in order to obtain external access to the object's data, for instance COM/DCOM (Component Model/Distributed Component Object Model), CORBA and RFC.
- **Interface Layer** – The Interface Layer describes the implementation and structure of a SAP Business Object and it defines the object's interface to the outside world.
- **Integrity Layer** – This layer represents the business logic of the Business Object. It comprises the business rules and constraints, which apply to a Business Object.
- **Business Object Kernel** –The core of a Business Object represents the object's inherent data.

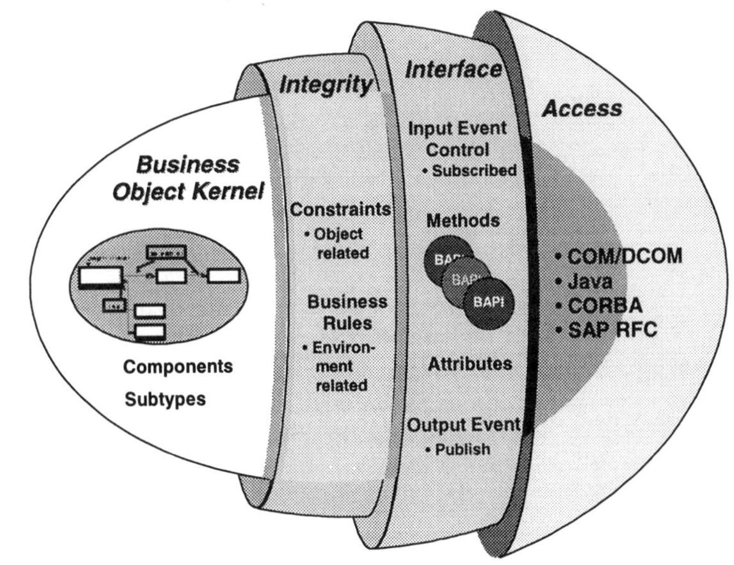

Figure 7-4: Business Object [SAP96l]

7.2.1 Access to the Business Objects

As was shown in figure 7-4, the interface layer functions as a 'buffer zone' between a Business Object's data and the applications and technologies that can be used to access it. To the outside, the SAP Business Object reveals only its interfaces, which consist of a set of clearly defined methods[38], which are also known as BAPIs. External applications can only access the Business Object via the object's methods.

An application program, which accesses a SAP Business Object and its data, only needs the information required to execute the methods. Therefore, an application programmer can invoke the methods of a Business Object without having to know or consider the object's underlying implementation and coding details. The set of methods associated with a Business Object represents the object's *behaviour*. When a method is executed on a Business Object, the method can change or retrieve information from the object's internal *state* of data. The previously introduced Business Object '**Material**' (see figure 7-5), has several methods. One of them is the BAPI **'Material.GetDetail'**, which provides the user with detailed data for a specified material.

7.2.2 Object Types and Object Instances

"Each individual Business Object belongs to a specific object class, depending on the nature and general characteristics of the object" [SAP97d]. In SAP terminology these object classes are called *object types*. For instance the individual materials used within an organisation are all part of the **'Material'** object type. "The object types are descriptions of the actual SAP Business Objects that can exist in R/3;" which means, "each individual SAP Business Object is a representation, or *instance*[39], of its object type" [SAP97d]. For example, the material 'Bicycle Frame is an instance of the object type "Material". When an instance of a SAP Business Object is used by an application program, the object's instance responds only to the set of characteristics and methods defined for its own object type. The following points define the SAP Business Object types:

---

38 Note that the methods of a Business Object are considered in a theoretical way. In other words how the methods are actually realised (BAPI, RFC, etc.) is secondary at the moment, this step will be done in chapter 8.

39 In the OO-Technology producing a more defined version of some objects by replacing variables with values (or other variables) is called instantiation. Instantiation is used in object-oriented programming, producing a particular object from its class template. This involves allocation of a structure with the types specified by the template and initialisation of instance variables with either default values or those provided by the class' constructor function [Boo94; WH97].

Object Type

Object Type – The object type describes basic data and includes information such as the unique name of the object type, its classification and data model. The following figure 7-5 visualises the structure of the Business Object "Material", and its corresponding interfaces, key fields, attributes, methods and events within the R/3 Business Navigator[40].

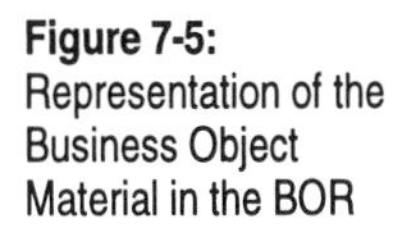

Figure 7-5: Representation of the Business Object Material in the BOR

Interfaces

Interfaces are groups of related methods associated with an object type. An interface type is a combination of attributes, methods and event definitions, which are used in a common R/3 context, for instance at the instantiation of a BO, like 'Material.Create' and its associated object interface IFCREATE.

An object type can support one or more interface types. The attributes, methods, and events already defined in the interface types are therefore available to the object type with interfaces and are also inherited by the respective sub-types [SAP96c]. The interface types are mainly used as internal references by the SAP BO developers.

Key Fields

The **Key Fields** determine the structure of an identifying key, which allows an application to access a specific instance of the

[40] The Business Navigator is the R/3 tool that lists the implemented R/3 Reference Models.

object type. The object type "Material" and the key field 'Material.Number' are examples for an object type and a corresponding key field. In other words a Key Field is the data type reference of a BO on the database level (Table: MARA, Key Field: MATNR).

Attributes

Attributes contain data about a Business Object, thus describing a particular object property. SAP divides their attributes into three types [SAP96c]:

One attribute represents a physical data field of the R/3 database.

A virtual attribute, which is computed on demand. Saving disk space and I/O activities.

References to other objects, like the material-attribute "Material.Document" that represents parts of the "document" object. In other words the "document" object is a component of the material object.

Methods

The above-mentioned attributes of BOs are manipulated by business transactions. These transactions are implemented into the **Methods** of a BO supporting the object-oriented concept of encapsulation. "The methods are invoked via the runtime component of the Business Object Repository transaction handler, which maps requests for method calls to the appropriate SAP Business Object method" [SAP96c]. A special set of BO methods was introduced with SAP's Business Application Programming Interfaces (BAPIs) which will be described in more detail in chapter 8.

Event

An **Event**, which is internally used in a R/3 system, indicates the occurrence of a status change of the assigned Business Object, "in a one-to-many relationship" [SAP96c]. An event handler handles changes or activities that occur at a specific object or other objects which are subscribed to that specific event. These events on BOs are mainly used in SAP's Business Workflow to coordinate the information flow between the application and the workflow management of the SAP system [SAP96c].

7.2.3 Inheritance and Polymorphism within R/3

One major advantage of the object-oriented technology is software reusability. Software reusability, in other words the reuse of software code, is achieved by deriving new object types from existing ones.

Supertype and Subtype

"When an object type is generated from an existing object type (see in figure 7-6), the new object type is called the *subtype* and the existing object type is called the *supertype*" [SAP97d].

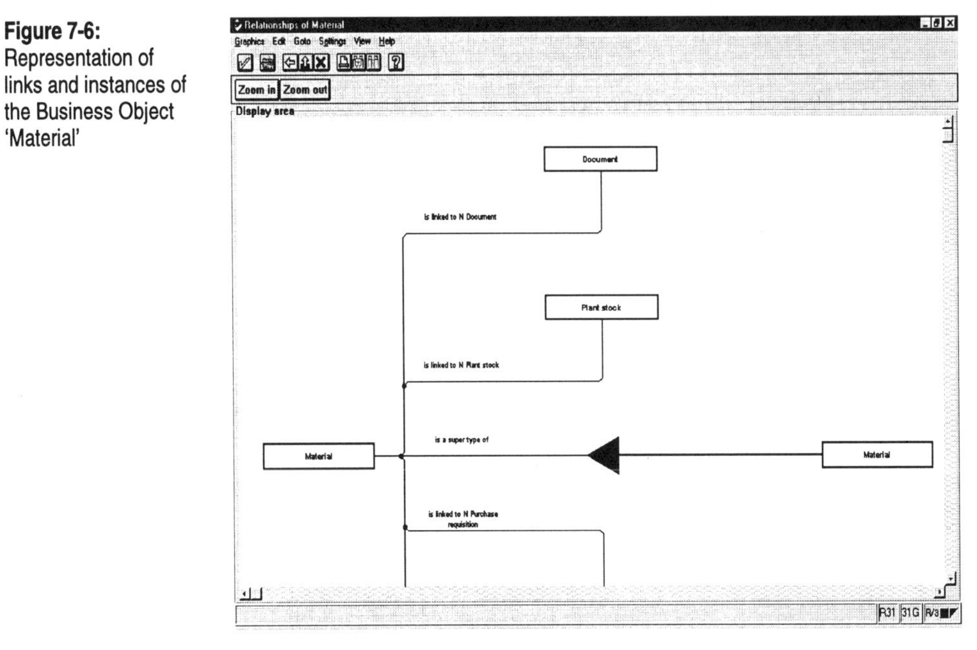

Figure 7-6: Representation of links and instances of the Business Object 'Material'

Inheritance

A subtype *inherits* all the properties and methods defined by the supertype from which it descends, but can also have additional properties and methods.

Polymorphism

A subtype may also implement a different behaviour for the methods, which were inherited, from its parent object. *Polymorphism* is the term that is used to describe the phenomenon that the same method shows a different type of behaviour when used in different Business Object types.

Figure 7-6 shows the realisation of the supertype and subtype assignment, which can be visualised within the R/3's ABAP/4 Workbench.

Note that this figure gives only an idea of how a printout of such a graphical representation of a SAP Business Object may look.

8 Business Application Programming Interfaces

As shown in chapter 7 the SAP Business Objects, which are held in the Business Object Repository (BOR), encapsulate their data. External access to the data is only possible by using their specific methods, which are called *Business Application Programming Interfaces* (BAPIs).

Why did SAP introduce the BAPIs?

In November 1996 SAP posted the first 100 BAPIs on their Internet site (www.sap-ag.de or www.sap.com). These were the first series of BAPIs and were the start of for a new way of interfacing programs with a R/3 system. The publishing of the BAPIs was important for the following reasons [FR96,1]:

1. BAPIs make R/3 more accessible. These High Level Interfaces enable the R/3 components and their Business Objects to 'talk' with one another. This is very important in supporting SAP's strategy to split the R/3 system up into individual components (a.k.a. componentisation). To give you an overview, BAPIs will replace the 45,000 non standardised RFCs used in the R/3 system today.
2. BAPIs allow competitors coexistence. Due to the non-standardised RFCs, SAP did not provide a solid interfacing basis for competitors and also partners.
3. BAPIs are SAP's commitment for an open R/3 interface support, which means that SAP will maintain the introduced BAPIs as long as possible.

SAP posts and updates all available BAPI in their so-called BAPI-Catalogue, which can be found on their homepage. The BAPI-Catalogue gives an overview of which interfaces are available but with no technical details. These can be obtained from SAP for a small fee for material and handling coverage.

Let us now take a closer look at the definition of the BAPIs and the technique, which stands behind them.

8.1 BAPI Definition

A BAPI is defined as a method of a SAP Business Object. For instance, when the 'Material' example we have already seen is considered, the detailed data of a Material Object can be accessed via the BAPI called 'Material.GetDetail'.

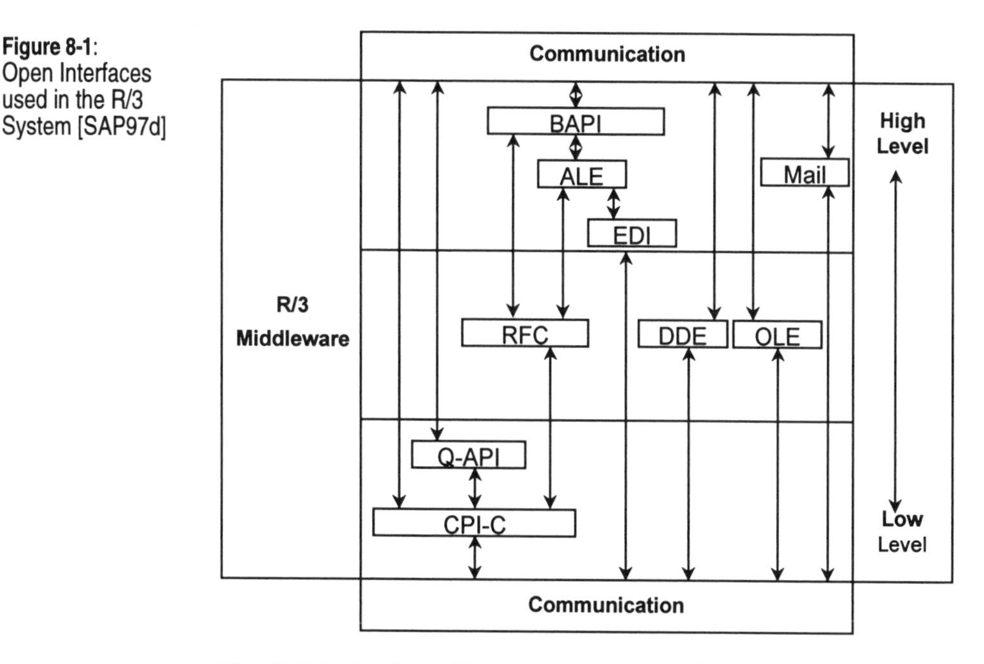

Figure 8-1: Open Interfaces used in the R/3 System [SAP97d]

The BAPIs in the R/3 system are currently implemented as so-called *Function Modules* (see figure 8-3), which are all held in the R/3 *Function Library.*

As already described, BAPIs are defined as API methods of SAP Objects, which are described and stored in the *Business Object Repository* (BOR). The BOR is the object-oriented repository in the R/3 System. It contains among other objects, SAP Business Objects and their methods. In the BOR, a Business Application Programming Interface (BAPI) is defined as an API method of a SAP Business Object. Thus defined, the BAPIs become standard with full guaranteed stability regarding their content and their interface.

BAPIs can be called up within the R/3 System, from external application systems and other programs. Examples of what BAPIs can be used for include:

- Distributed R/3 scenarios using Application Link Enabling (ALE),
- Connecting R/3 Systems to the Internet using the *Internet Application Components* (IACs),
- Workflow applications that extend beyond system boundaries,
- Customers' and partners' own developments,

- Connections to non-SAP software,
- Connections to legacy systems and
- Third party front-ends to R/3 Systems.

Generally we can say, that SAP is doing the right thing by introducing an industry-wide standard with their BAPIs. One of the main advantages in introducing them is an internal one. BAPIs are supposed to be an API wrapper around the R/3 system so that SAP is more flexible against any technology changes, which can occur.

The technology on which BAPIs are based on will be described in the following sections of this chapter.

8.2 Function Modules in the Function Library

The R/3 "Function Modules are external subroutines that are stored centrally in the Function Library and can be called from any ABAP/4 program" [SAP97e] or an external program via the Remote Function Call (RFC) protocol. In contrast to ordinary subroutines, Function Modules have a uniquely defined interface [SAP97e].

The BAPIs, which are accessible from an external R/3 or non-R/3 system via RFC, for example 'Material.GetDetail', are marked in the Business Navigator with a 'green dot' (see figure 8-2). The rest of the methods are either not implemented (as a RFC function) or are used internally.

Figure 8-2: Representation of the Business Object Material

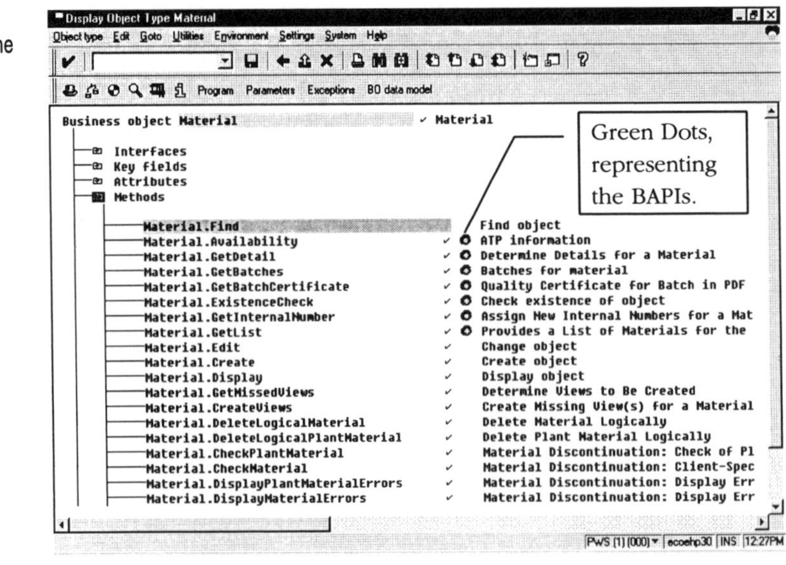

"The Function Library manages Function Modules", within the ABAP/4 Development Workbench, "and provides useful search functions for retrieving information on them" [SAP97e]. The Function Library supports the user in the creation of new Function Modules or in maintaining existing routines.

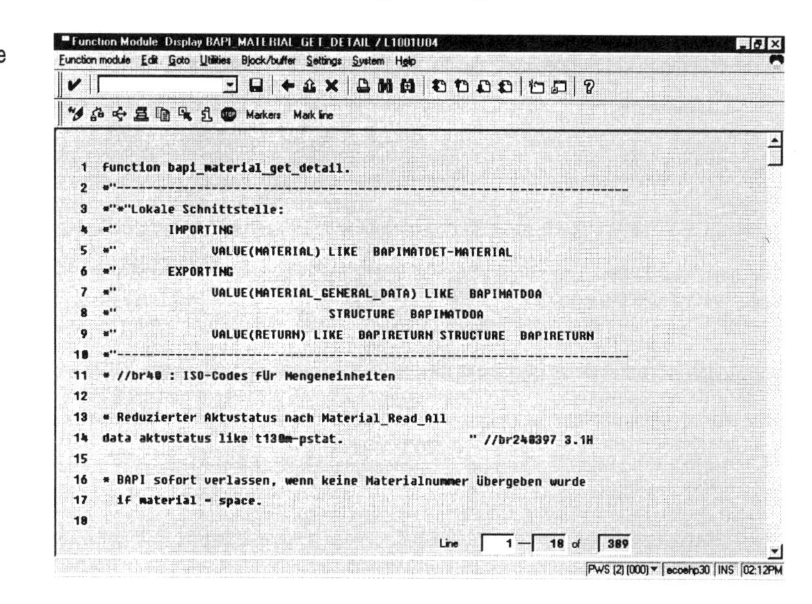

Figure 8-3: The Function Module for BAPI: 'Material.GetDetail'

8.3 The BAPI Interface

To use a BAPI, an application program only needs to call the method with the right interface parameters, according to the BAPI's definition. Therefore, when a BAPI call is included in an application program, the application developer only needs to provide the appropriate interface information to use the BAPI correctly.

A BAPI interface is defined by its [SAP97d]:

1. **Importing parameters** – They contain data, which has to be transferred from the calling application to the BAPI.
2. **Exporting parameters** – They supply the data, respectively the result, which has to be transferred from the BAPI back to the calling program.
3. **Table parameters** for both the importing and exporting data.

4. **Exceptions** –They indicate that an error has occurred during the BAPI call and is also responsible for managing the necessary exception handling.

Figure 8-4 shows the interface definition of the BAPI 'Material.GetDetail' represented in the R/3 system. The BAPIs are a high level interface within the R/3 Middleware which takes care of the various interfacing technologies and protocols, which are possible in a R/3 environment.

The R/3 Middleware (see figure 8-1), where the RFC technology belongs to, offers different communication interfaces on different levels to the application programmer.

Figure 8-4: Representation of the Material BAPIs

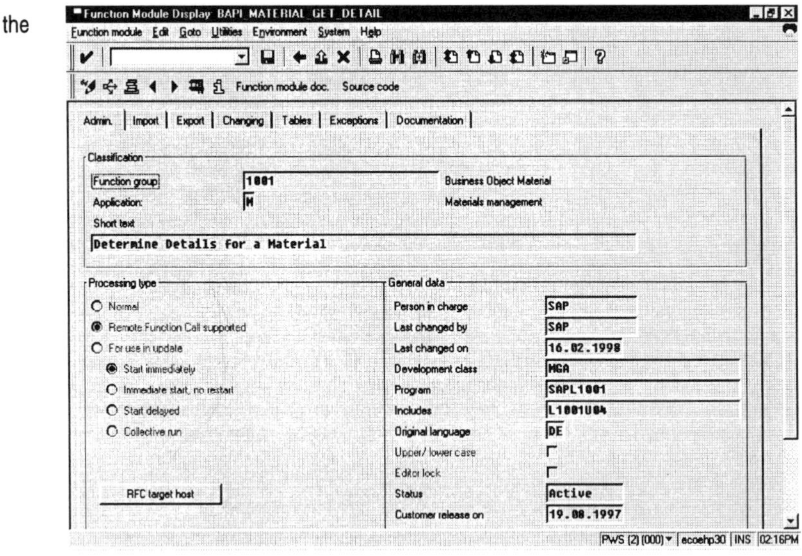

The communication services of the higher levels themselves utilise some of the services of the lower levels for their own functions.

With the R/3 Release 3.1G or 3.1H, BAPIs are only based on SAP's RFC-technology but with Release 4.x, ALE- technology, based on tRFC-technology, will be used additionally. The following sections will cover both protocols in more detail.

8.4 The Remote Function Call (RFC)

"A remote function call is a call to a function module running in a system different from the caller's. The remote function can also be called from within the same system (as a remote call), but usually caller and callee will be in different systems.

In the SAP System, the ability to call remote functions is provided by the *Remote Function Call* (RFC) interface. This interface allows remote calls between two SAP Systems (R/3 or R/2), or between a SAP system and a non-SAP System" [SAP96j].

The tool for developing RFC programs is the SAP RFC SDK (Software Development Kit), which contains executables, libraries, 'INCLUDE' files as well as ABAP/4 and C programs for the various hardware platforms. R/3 has a built-in *RFC Generator* which utilises some of the RFC SDK functionalities: This RFC Generator is able to create fully functional programs such as relevant C or Visual Basic programs for an existing function module with a fully defined interface in the client or server version.

The RFC technique can be used to address programs or services, running on a R/3 application- or front-end-server.

Figure 8-5 visualises a possible implementation of a RFC Client/Server scenario including the relevant RFC functions used to establish a connection.

Figure 8-5: RFC Client Program [SAP97d]

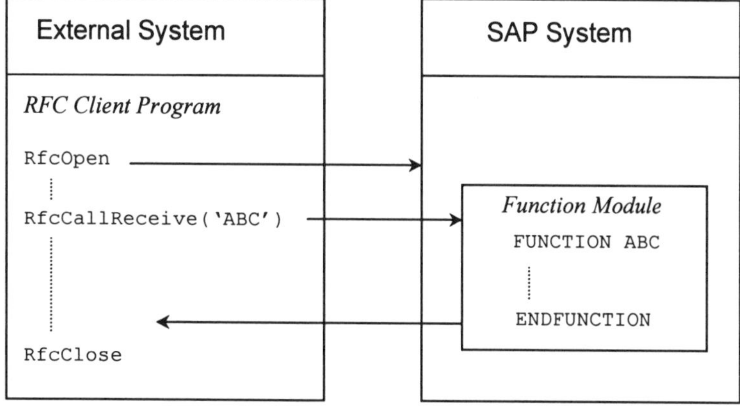

The RFC technology is structured similarly to the Remote Procedure Calls, for example used in the UNIX operating system, to establish Client/Server programs [SAP96j].

So how does this technique work? Firstly, the external system has to establish a connection to the R/3 System with the **RfcOpen** command. Secondly, after the successful creation, the appropriate R/3 Function Module can be called. The call of a function is achieved through the **RfcCallReceive** command, which executes in this example the ABC function and retrieves the returning R/3 results. The last step in this sequence is responsible for closing the connection, which is done through the **RfcClose** command.

A 'little' weakness of RFC

An RFC client program (ABAP/4 or external program) will usually repeat the call of an RFC function if a network error (CPI-C error) is returned by this call. If it is a network error caused by calling an RFC function, this action has to be repeated. However, if this network error occurred during or at the end of the execution of an RFC function, the RFC function may already be executed. In both cases, the RFC component in SAP Systems or in the RFC library will have the same CPI-C error code from the CPI-C layer. It is not advisable to repeatedly call up this function, because the required RFC function will be executed one more time and would leave the R/3 database in an inconsistent state.

Being aware of this problem, SAP introduced the transactional RFC (tRFC), to ensure safe transactions. These tRFCs are described in the following section.

8.5 The Transactional RFC (tRFC)

From Release 3.0 onwards, data can be transferred between two R/3 systems reliably and safely via *transactional RFC* (tRFC). This ensures that the called Function Module executes exactly once in the RFC Server System.

Moreover, the RFC server system or the RFC server program does not have to be available at the time when the RFC client program is executing tRFC.

The exact execution of one Function Module is achieved through a unique *Transaction Identifier* (TID), which is exchanged between the Client and the R/3-Server program to ensure a safe transaction. In other words both the tRFC client and the tRFC server program have to manage the TIDs themselves for checking and executing the requested tRFC functions exactly once.

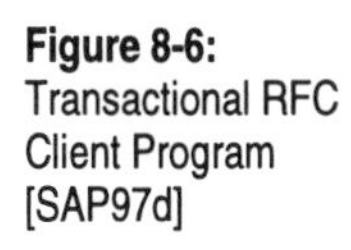

Figure 8-6: Transactional RFC Client Program [SAP97d]

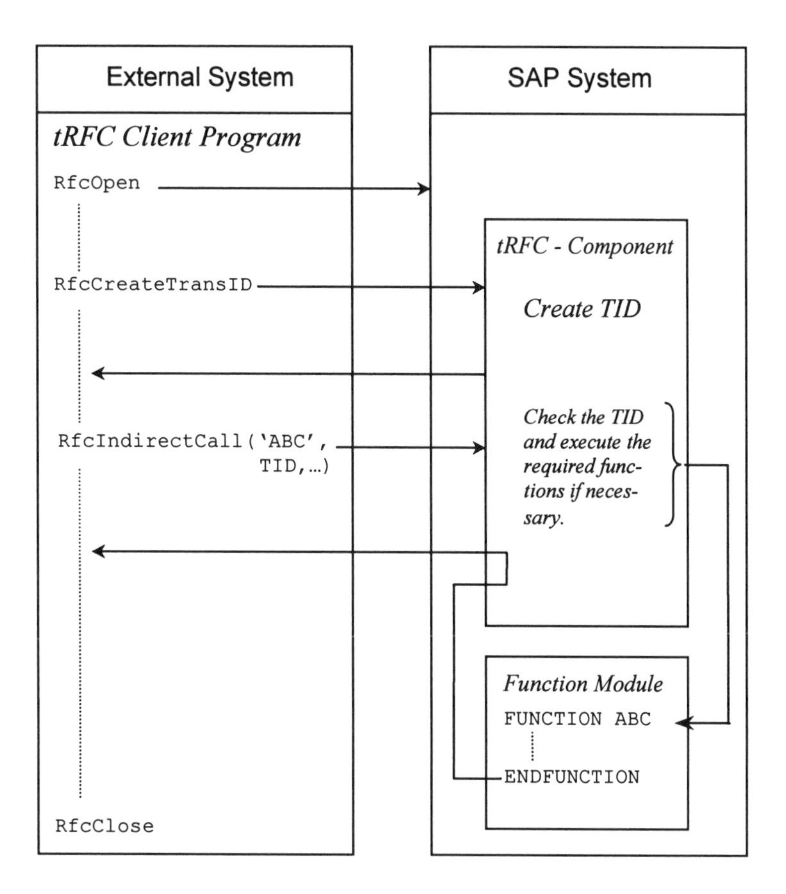

9 Application Link Enabling (ALE)

Distributed applications may have different drivers but one of the major challenges most geographically dispersed companies share is that they are facing an increased pressure to provide better products and services, at lower costs, and with increased shareholder value.

9.1 Definition of ALE

ALE makes distributed systems possible by allowing users to set up application modules and databases at different locations. In other words, SAP's ALE concept removes the constraints of a single database by distributing databases among different sites.

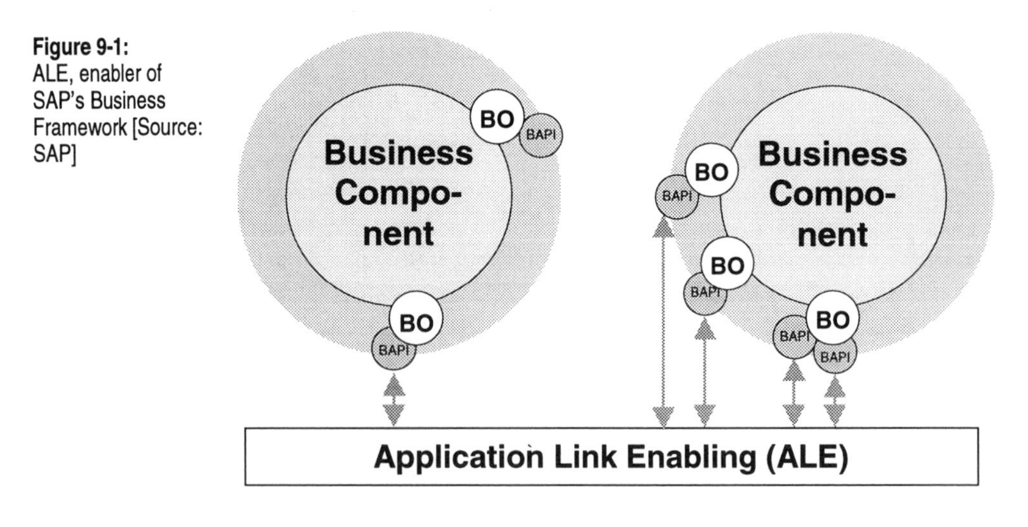

Figure 9-1: ALE, enabler of SAP's Business Framework [Source: SAP]

ALE is a set of:

- Tools,
- Programs and
- Data Definitions,

that provides the mechanism for distributing SAP functionality and data across multiple systems. Furthermore ALE allows the user to distribute:

- Control or Customising Data is transferred to organise the corresponding systems and access files such as user profiles and business organisation data. Data about the business organisation can include the definition of configuration objects, enterprise structure, global settings etc.
- Master Data is important data that enables the local systems to determine whether a certain master data object already exists in the system or not [SAP96k]. This data, which is transferred completely or as an update on a previous complete transmission to reduce the quantity of data, is messaging between the ALE systems [SAP96k]. Master data is for example material, customer, vendor master etc.
- Transaction Data which can be exchanged between two ALE systems includes customer orders, purchase orders delivery notices and bills [SAP96h].

All this is supported by the predefined ALE Business Scenarios, approximately 100, which are ready to use.

Sometimes ALE is compared to the concept of Distributed Database and Database Replication, but these concepts are completely different!

The concept of Distributed Databases belongs to the database replication technology, which means that at a specific point in time or based on specified events, data is replicated from one database to another. The emphasis here lies on data exchange at database level, whereby the ALE approach of distributing data is triggered and controlled by an R/3 application not by the database itself.

SAP's ALE concept was not a totally new approach, rather SAP based ALE on EDI's functionality.

What is EDI?

EDI stands for Electronic Data Interchange and is used in the business world to send and receive purchase orders, sales orders, invoices and other business documents via one of the agreed EDI standards e.g. EDIFACT, ANSI. The intention of EDI is to provide a mechanism for sending messages between computers, so that they can read them regardless of the software they are running on. SAP adopted the already existing EDI standards to use them for their ALE technology. Both ALE and EDI envision sending and receiving data by so-called message-types.

Difference between ALE and EDI?

Since the ALE concept is presently based on the EDI interface, it makes no difference to the application whether data communication is carried out using an EDI scenario or an ALE scenario. The application can recognise whether data is to be sent and generates a master IDoc. The setting in outbound processing determines the route via which the IDoc is to be communicated; for example, if an ALE port or an EDI port is maintained, and how the partner profiles are set. The main difference between EDI and ALE are:

- A transmission, which uses ALE, is carried out directly by calling up a function module (tRFC-based) in the remote system.
- If EDI is used, an EDI subsystem has to be inserted between the broadcasting and the receiving system that handles the transmission.

ALE Process Flow

There are four different ways of using ALE:

1. SAP R/3 to SAP R/3 – This approach is the traditional definition of ALE and is the most commonly used, when we speak about a distributed SAP installation.
2. SAP R/3 to Non-SAP System – This approach uses IDocs to send the data to an external (third-party) system. The advantage is that the customer does not have to write code for the data extraction. The disadvantage is the need for a conversion tool (Middleware) or inbound-customer coding to import the data into the Non-SAP system.
3. Non-SAP System to SAP R/3 – This connection is used to load data into the R/3 system. The use of outbound customer coding or a Middleware software is needed to put the data into the IDoc structure of the R/3 system.
4. SAP R/3 to EDI – Due to the similarity and that both technologies are implemented within a SAP R/3 system. The setup is almost the same as a SAP R/3 to SAP R/3 scenario.

9.2 Implementation Benefits

The questions that arise now are, what are the implementation benefits for using ALE for distributing a company's R/3 System over different R/3 instances?

When speaking of Implementation Benefits, it should be clear that the ALE approach is now compared to a common R/3 setup, which is one single R/3 instance based on one database server.

System Performance

The first point, which has to be mentioned is, that with an ALE implementation possible bottlenecks on a system can be removed. For instance, the heavy workload that is to say transaction load on a central system can be distributed to multiple R/3 systems.

High System Availability

High System Availability is necessary for globally working organisations. ALE enables them to distribute data to multiple sites so that system maintenance work (backup, upgrade etc.) can be done on systems over night when they are not in use.

SAP R/3 Release Coordination

In a single instance environment it may be necessary to get the latest R/3 Release due to changes in one component. Let us assume that changes in tax laws forces a company to change from one SAP Release to the succeeding version. Through the componentization of a R/3 system, a more efficient release coordination can be achieved through ALE. This also means, that if R/3 applications are distributed on different systems, different R/3 releases can be obtained, due to the compatibility of the ALE Business Scenarios. The future plans of SAP are to split the R/3 system into individual components and the ALE technology including the BAPIs will be used as the glue for connecting them.

Very Large Databases (VLDB)

A very large database of a R/3 system can also cause the company to distribute their system onto multiple machines, so that this performance bottleneck can be resolved.

Centralized Consolidation and Business Information Warehouse

More and more companies would like to have a centralized Information System (IS) where they can do certain reporting on consolidated data as a single point of all enterprise information. Common R/3 systems (also called Online Transaction Processing (OLTP) systems) are supporting this reporting functionality with their IS for logistic data, financial data etc. Due to the intensive needs of system performance for these calculations special, Data Warehouse (DW) systems have been invented. These systems are designed to process intensive queries on consolidated data which they have received from external source systems. SAP's notion of a Data Warehouse is called Business Information Warehouse (BW) – also known as Online Analytical Processing (OLAP) systems.

The major advantage of SAP's BW is that it has the same meta data repository of a common R/3 system so that the source data can easily be transferred from the source system to the DW. ALE is used in this scenario to transport the extracted data from a source OLTP system to the OLAP system of the BW.

Generally speaking as SAP adds Internet, workflow and business object interfacing technology to their R/3 system, the demand for a sophisticated Middleware technology increases as well. The solution for this can be found in ALE. Let us have a closer look now to the technical side of ALE.

9.3 ALE Layers

ALE provides three services for distributing applications and also the tools to tailor them to the requirements of an organisation. The three layers are:

1. Application Services – These services are not specific to ALE but this is where the individual SAP applications (SD, FI, CO etc) generate their data and documents (e.g. generating an invoice, posting an order)
2. Distribution Services – Here, the actual distribution of the data and documents takes place. This service layer determines where the recipient resides and the corresponding data is filtered and formatted into the IDoc structure.
3. Communication Services – This service layer is responsible for the distribution of the ALE IDoc from the sending to the receiving system.

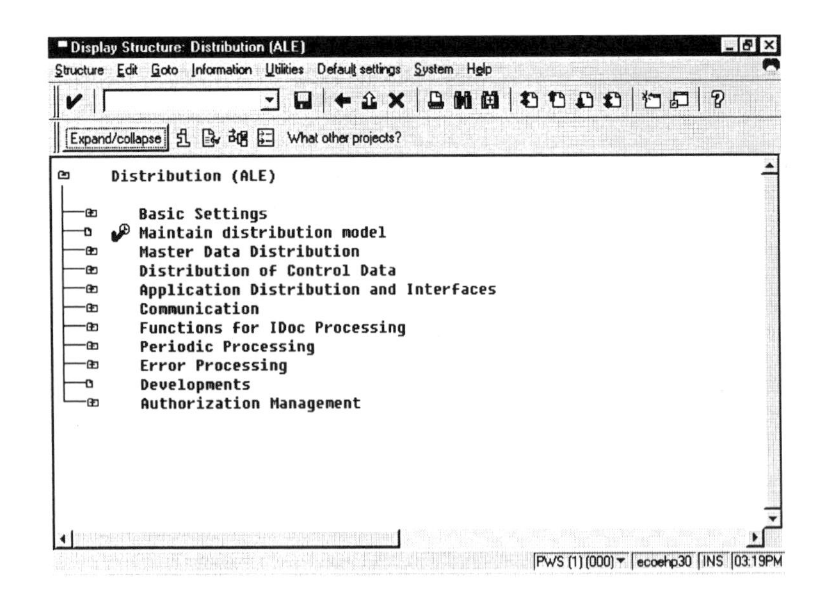

Figure 9-2: ALE Customising in R/3

As mentioned before, the IDoc is the main component of the ALE and EDI interface functionality. The IDoc is structured into three record types:

- Control Record – This shows the header information that uniquely identifies the IDoc.
- Data Record – Its responsibility is to store the application data related to the associated message type.
- Status Record –This information is used for audit trails to see where the IDoc came from, where it went and its statuses through the life of an IDoc.

ALE Customising

An application in R/3 must be configured to allow the creation of IDocs. The following picture shows the part of the SAP R/3 IMG (Implementation Guide), where the customising parameters 1have to be maintained to be able to send IDocs from one system another.

Figure 9-3: ALE Layer within R/3

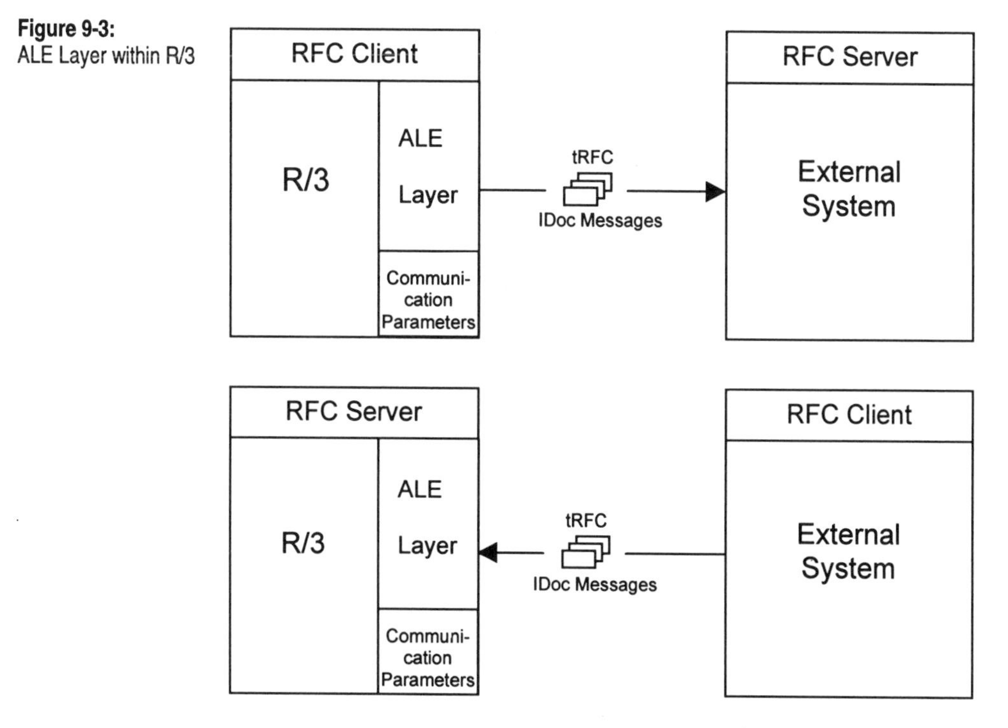

When data is transferred to another system, a master IDoc is generated. The IDoc is transferred to the ALE layer, where it is formatted for sending and then communicated. After the transmission has occurred, the IDoc is imported into the target system where it initiates inbound processing. Depending on the configuration settings, the data is processed either automatically or manually. Outbound or Inbound processing can be done individually for every IDoc or in bulk processing. Communication between the distributed systems is carried out with IDocs asynchronously. In most cases, routing of the data is initiated immediately after the update in the application.

9.4 ALE and BAPIs

Starting with the R/3 Release 4.x, SAP combined the ALE and BAPIs as their major interfacing technology.

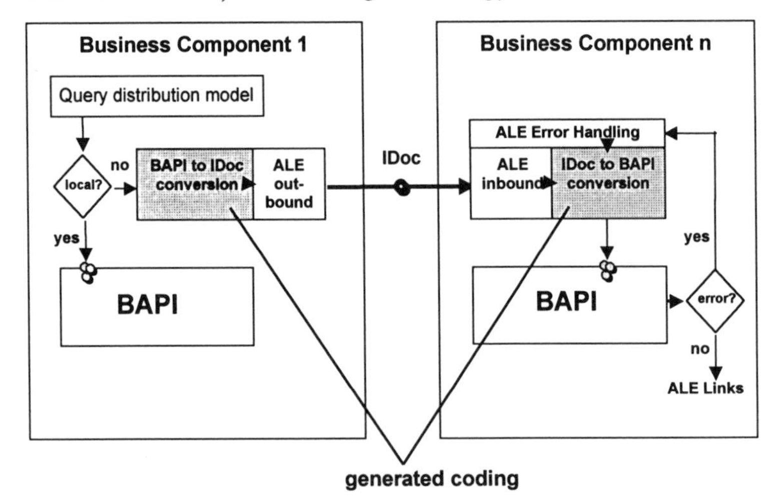

Figure 9-4: BAPI invocation by ALE [Source SAP]

The usage of BAPIs within the ALE scenario has the following advantages:

- Integration of an object-oriented approach/technology.
- Only one interface has to be maintained.
- Programming errors can be avoided through the automatic code generation.

Generally speaking the combination of both technologies closes the existing gap that BAPIs could not handle asynchronous communications. This combination enables BAPIs to use the tRFCs so that they can also be invoked asynchronously. Additionally, when business functions are distributed, they can use the already existing methods such as the ALE error handling and ALE audit functionality.

10 Prototype - Design

Anyone who sells for a living knows that taking care of your customer is the only way to get future orders. Taking care of your customer means providing the customer with solutions to his/her specific business problems, offering information to help him/her in the job, providing the customer with opportunities to learn more about the company they buy from and generally consider the customer's problem his own problem. This was true in the early 1900's and will continue to be true well into the 21st century.

10.1 Customer centric services

The purpose of customer-centric services is to completely integrate our processes and technology around the customer. This does not simply mean keeping track of a customer name and address database, or making sure a valid quote is assigned to a valid customer, these activities are developed to service an internal corporate/administrative requirement.

It also does not mean giving equipment/products away but binding the customer closer to our company and developing a strategic way of solving problems with him/her now and in the future.

> *"A single most important thing to remember about any enterprise is that results exist only on the outside. The result of a business is a satisfied customer. The result of a hospital is a healed patient. The result of a school is a student who has learned something and puts it to work 10 years later. Inside an enterprise are only costs."*
>
> *The New Realities [Peter F. Drucker]*

Account Managers and Sales Representatives (reps) have become comfortable with their laptops and the stand-alone applications on their laptops. Therefore, it makes sense for the developer to write systems which are seamlessly delivered to the laptop, creating a virtual environment where sales, marketing and customer care become one integrated activity. It also empowers the salesforce to respond immediately to daily market changes.

Sales reps who support their customers are evaluated by four simple principles – price, quality, speed and service. Price and quality are determined by the product they sell and the factory that produces it. Speed and service are manifested by the work habits of the rep and the tools available for the rep to support his/her customer [Mai95].

Speed and service, however, are modifiable characteristics. As our customers become more computer literate, definitions of 'fast' become more rigorous. The speed with which computers allow us to serve and support our customers is the common thread in each of the two 'encounters' at the start of this prototype. To exceed our market's expectations of speed, customer requirements must be tightly integrated into the company's system – one in which customer data, orders and preferences are loosely monitored and satisfied. The goal goes even further – a strategic IT environment has to be built, which enables the user to monitor the market, customers and the competition so that he/she can pro-act to any changes rather than react. This way he/she may be able to keep our competitors from gaining market share.

10.2 The Prototype

The idea behind this prototype is to develop a *Technology Enabled Sales* (TES) product, that is to say application that uses the strength of Lotus Notes as a groupware product that enables mobile computing of a company's sales force (see figure 10-1).

Figure 10-1: The TES Prototype Scenario

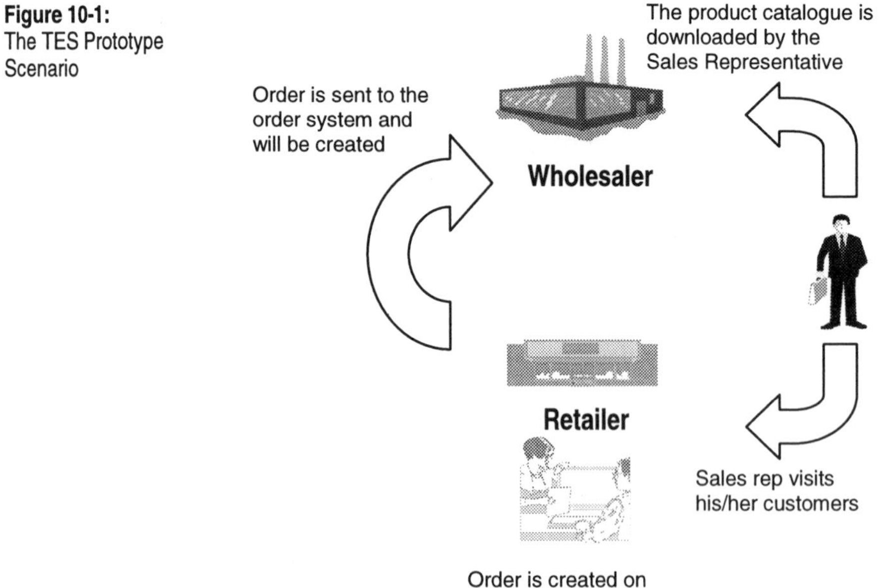

The R/3 System on the other hand reflects the core system of the company that stores and handles all essential data, necessary for the company's day-to-day business. The scenario from this system should support the sales force of a bicycle-wholesaler where the individual sales rep has to visit the customer regularly to get the necessary leads. The following sections of this chapter describe the design of this prototype in more detail.

TES - Prototype Scenario

As mentioned above, this chapter will describe the Lotus Notes - SAP R/3 prototype's design, which will support sales reps, working for a fictitious bicycle wholesaler company. The prototype will support sales reps during their regular customer visits and also during their work in the office to keep track of customer orders. The TES application will cover the following tasks:

- Product Catalogue
- Customer Address Management
- Order Entry

TES - Architecture

The following figure shows the IT environment, in which the TES application can be used.

Figure 10-2: Envisioned TES Environment

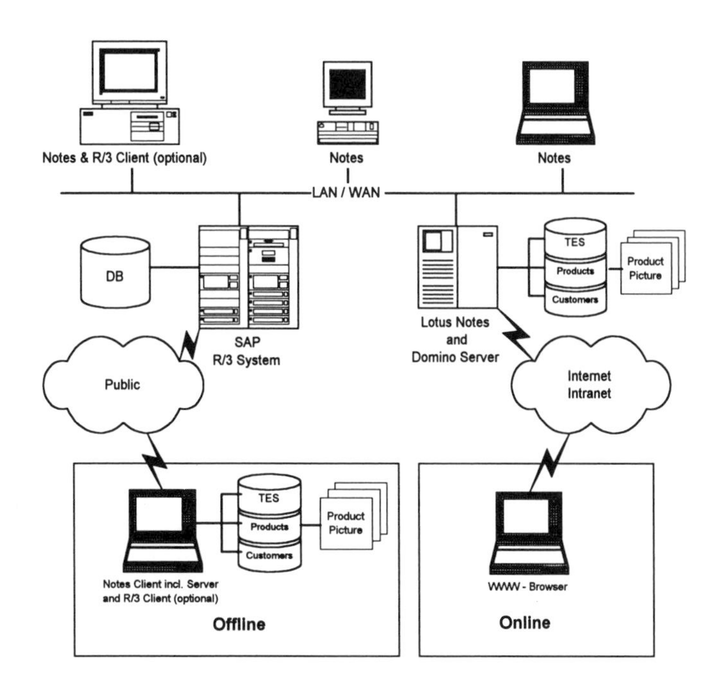

The TES user is able to do his/her work in different ways.

1. The sales rep creates the customer orders offline on his/her laptop. The TES client and server resides on his/her machine. At the end of the day he/she is then able to connect to the R/3 system, via telephone line (from home or a hotel) or via a LAN/WAN connection that exchanges data between the TES system (Lotus Notes 4.5) and the R/3 System (Release 3.1G). This scenario will be realised in the TES prototype.
2. The second possible way is to connect to the system online via a WWW-Browser. This is supported by the Internet/Intranet capability of Lotus Notes, respectively its Domino server.

The environment setup for developing the prototype is shown in the figure 10-3.

As shown in figure 10-3, the TES application, which resides on a desktop computer, is divided into the following three Lotus Notes databases.

1. TES Database – handles the main menu and so enables navigation in the TES application.
2. Products Database – contains all R/3 product data which can be downloaded from the R/3 system into the TES application. Moreover, it provides further on visual images of each product, which are linked and stored locally on the user's machine.
3. Customers Database – manages the R/3 customer data. New customers can be created and updated from TES to the R/3 system.

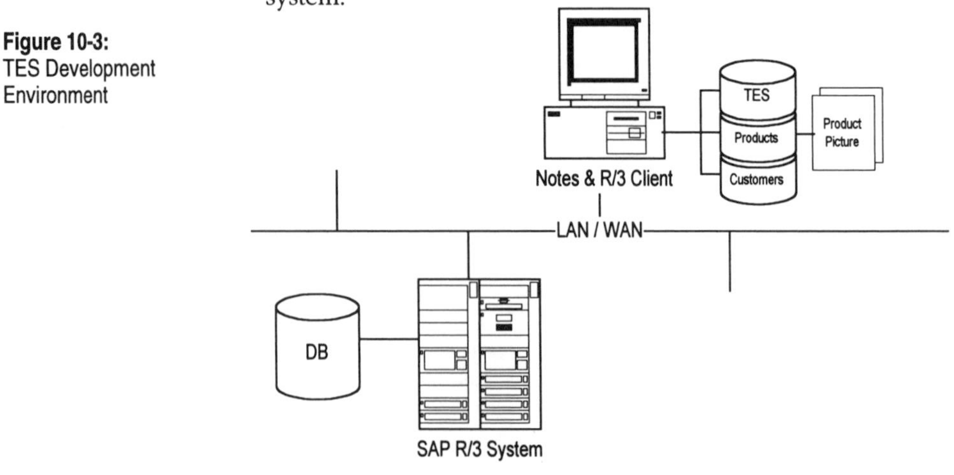

Figure 10-3: TES Development Environment

The following section describes how the TES prototype will work from a theoretical perspective. It will visualise the data flow within the different Notes databases as well as the flow between Lotus Notes and the R/3 System.

In addition to that this section will give a semiformal description of the functions, realised in the various submenus of the TES system.

10.3 TES Application Overview

Figure 10-4: TES Application Overview

Figure 10-4 visualises the structure of the TES prototype application. The Notes database 'Bicycle Shop-TES Main Menu' contains and handles the main menu navigation of the whole application.

The Notes database 'Product Catalog' contains all main product information retrieved from the SAP R/3 system and the necessary R/3 link information.

The Notes database 'Customer Orders' contains all main customer data (see menu 9 in figure 10-4) retrieved from the SAP R/3 system and also allows new customers to be created locally on Lotus Notes and to transfer them to the R/3 system. This database contains a so-called 'Shopping Basket' section later on where all selected products are stored temporarily before a customer order will be created.

10.3.1 Main Menu

Figure 10-5: Structure of TES' Main Menu

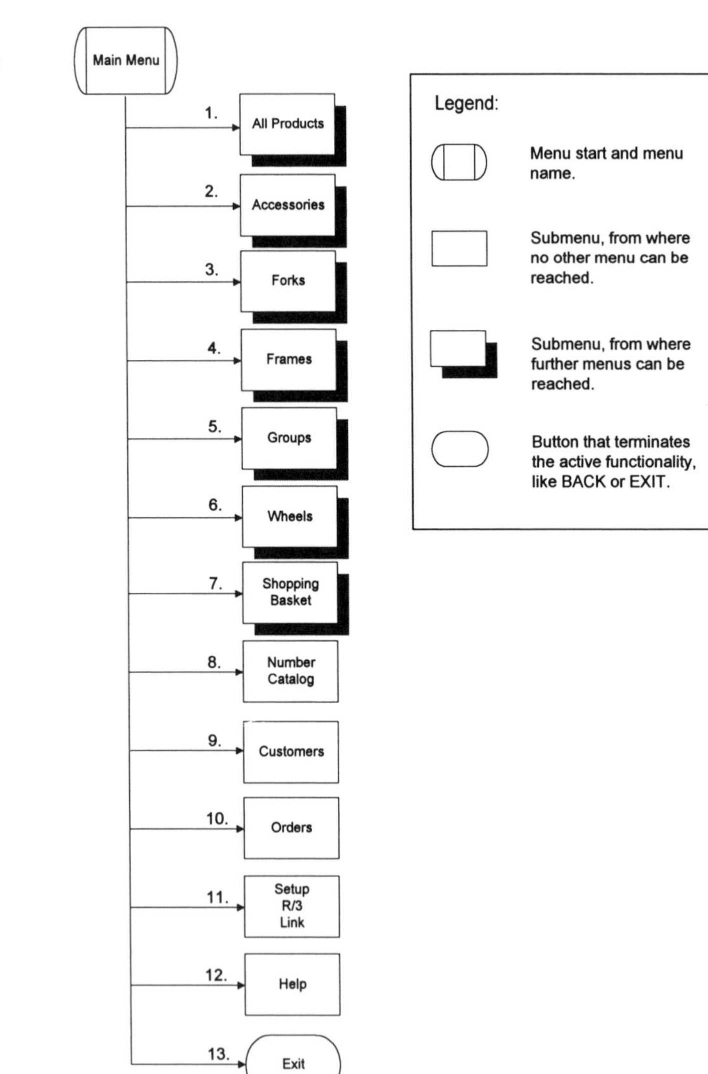

The main menu provides the user with an overview of the whole application functionality. Points 1 to 6 are the core of the product section, where either all products or the different product categories can be viewed, selected and put into the 'Shopping Basket', shown under point 7.

Points 8 to 11 concern the management of additional information, needed for the TES product, like customer data or information about already created orders.

Submenu 11 deals with the 'Setup R/3 Link', where all the necessary information for a successful synchronous Lotus Notes-to-R/3 connection is managed. Submenu 12 and 13 are self-explanatory.

10.3.2 Submenu – All Products

Figure 10-6: Submenu structure – All Products

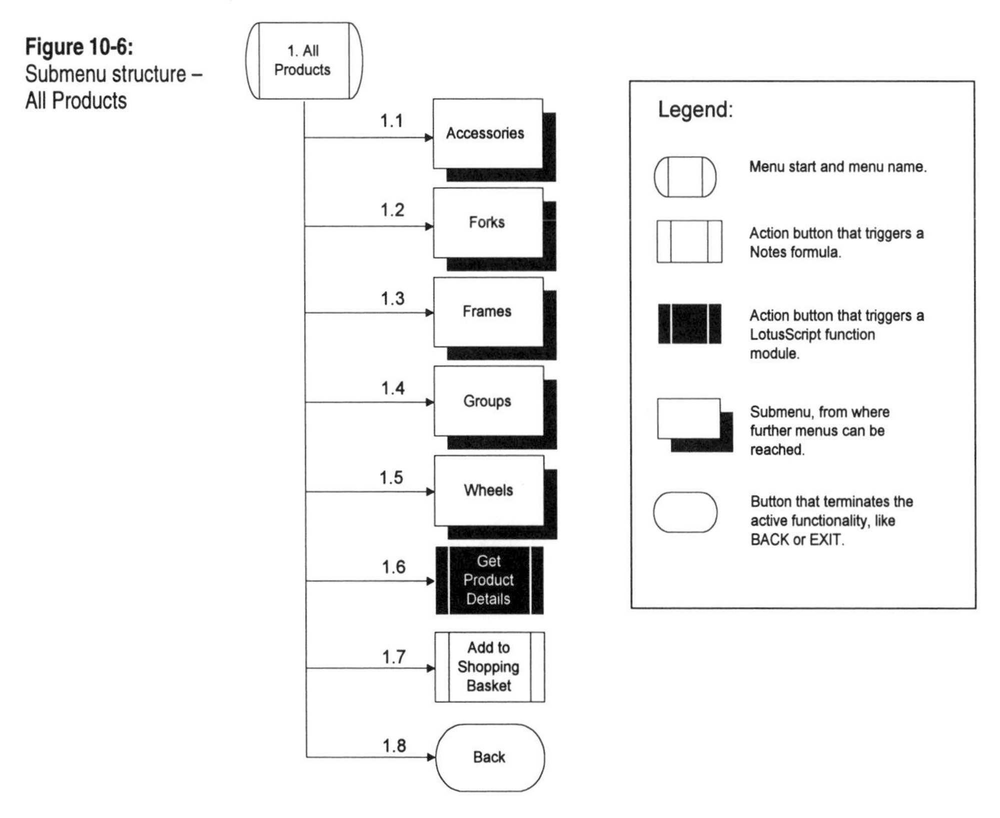

Note that due to similarity reasons the above shown submenu is used as a representative for the submenus 2 to 6 of figure 10-4.

All menus concerning the various material views are visualised in this submenu. These support the user in product selection.

This is also the application level where R/3 product data is retrieved with the help of the *Action Button*[41] 'Get Product Details'. The functionality behind this button uses the information stored in the submenu 'Number Catalog'.

After a product selection, the user is able to copy the selected products into the virtual 'Shopping Basket' through the Action Button 'Add to Shopping Basket'.

This button (see point 1.6 in figure 10-6) is taken as an example to describe the main functionality, implemented to retrieve or send data between the TES and the R/3 system.

The flowchart in figure 10-7 shows the structure of the LotusScript module that is implemented behind these Action Buttons.

Function	**Description**
<Button XYZ>	Stands for the CLICK-event of the button that triggers the LotusScript module XYZ.
OpenConnection	This function is responsible for opening a synchronous connection to the specified R/3 system. The setup is done in the submenu 'Setup R/3 Link' which saves the system settings in the **notes.ini** file. The OpenConnection function will then take the chosen R/3 system settings from the notes.ini file and opens a synchronous connection to exchange data between the TES and the R/3 system.
<Function XYZ>	Represents the module, which incorporates the main functionality of the chosen task.
CloseConnection	This function closes the synchronous connection between the TES client and the R/3 system.

[41] Action Buttons enable a user to perform tasks that are associated to the current view. Action Buttons are located at the action bar area, which is displayed just below of the Lotus Notes SmartIcon palette. Action Buttons can trigger funcitonalites, which are programmed with formulas or with LotusScript.

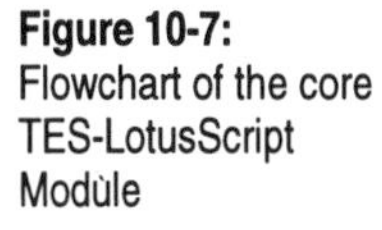

Figure 10-7:
Flowchart of the core TES-LotusScript Modùle

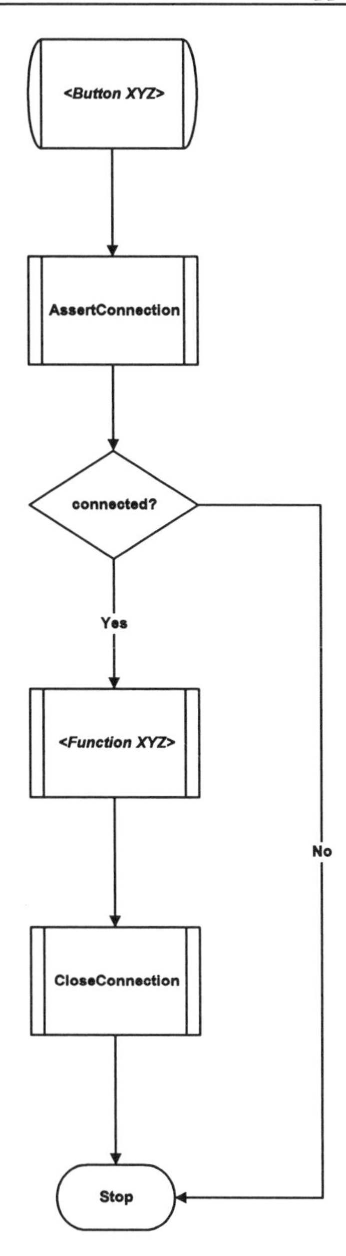

Each time a 'black'-button (referring to the different submenus) is pressed, the application runs through this procedure where each one looks similar. A more detailed technical description of the various LotusScript functions and their functionality will be made in the next chapter.

10.3.3 Submenu – Shopping Basket

Figure 10-8: Submenu structure Shopping Basket

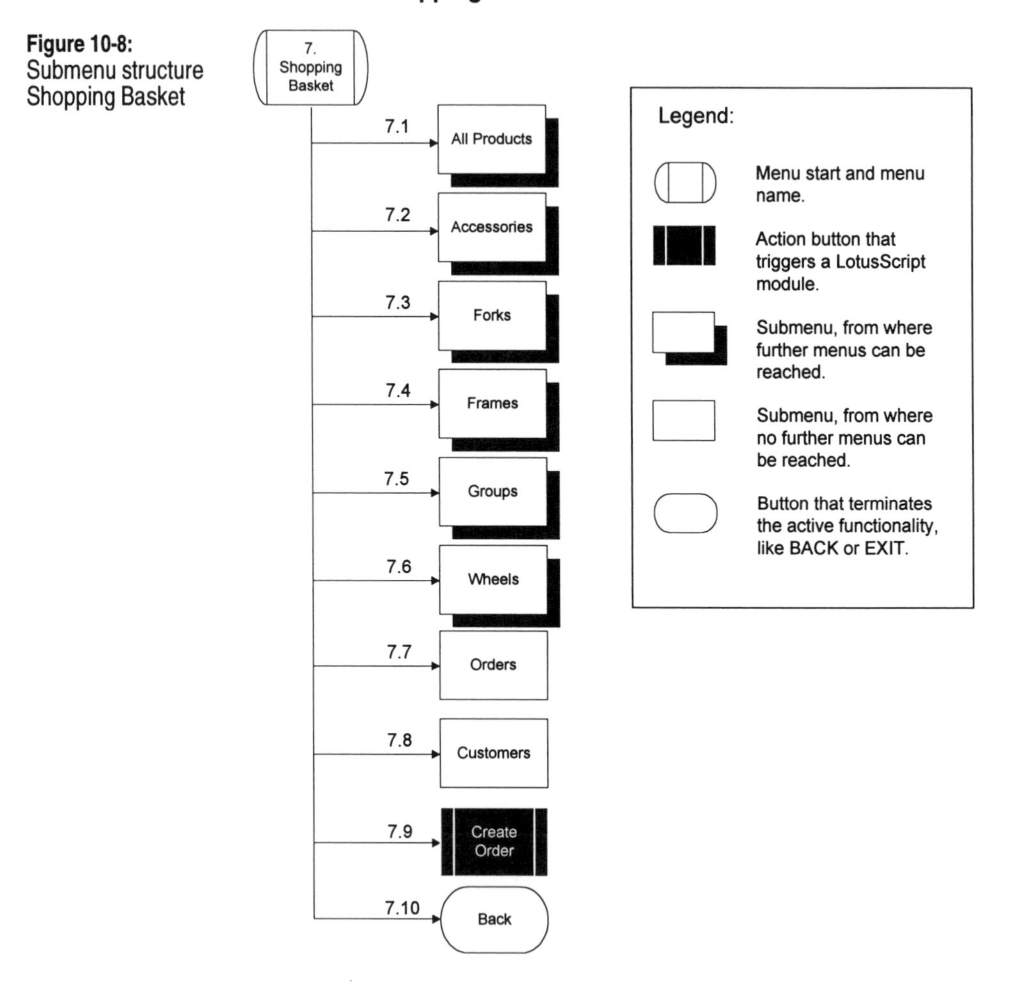

The menu 'Shopping Basket' holds the selected products. This selection is then the base for a customer order.

This customer order can be created through the Action Button 'Create Order' (see button 7.9 in figure 10-8). This order fills the order form with the selection and deletes the Shopping Basket entries.

The user is also able to switch from here to the 'Orders' and 'Customers' menu.

10.3.4 Submenu – Number Catalogue

Figure 10-9: Submenu Structure – The Number Catalogue

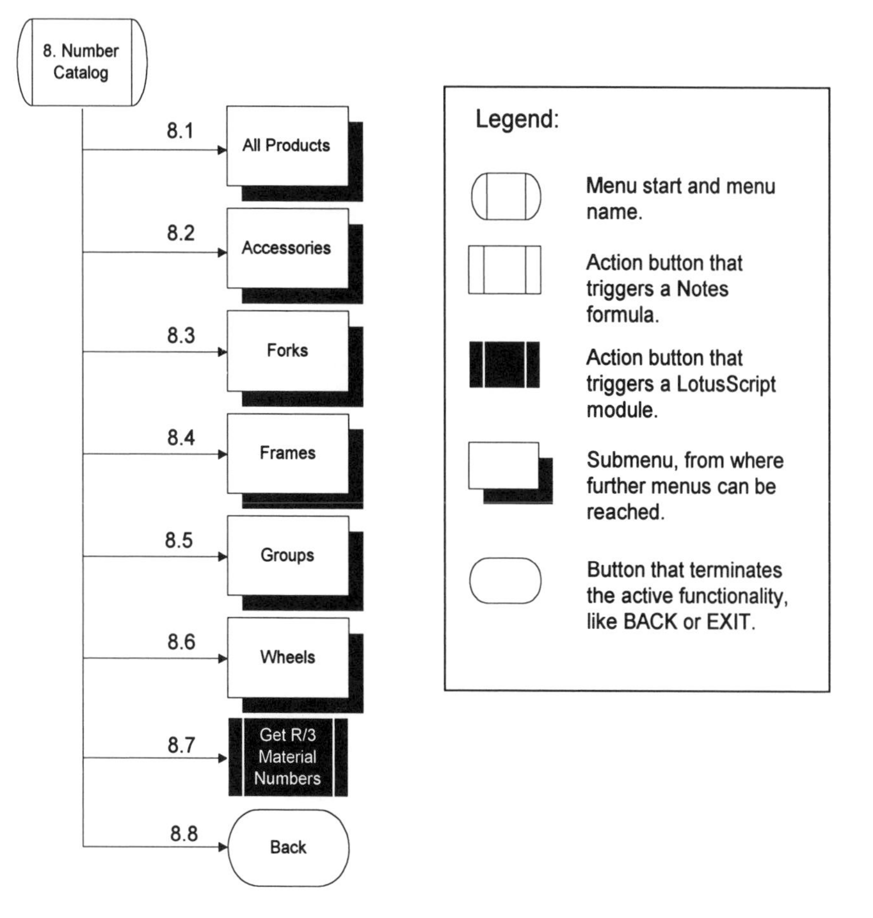

The submenu 'Number Catalog' holds all possible R/3 material numbers. These are necessary for the functionality behind the 'Get Product Details' button in the submenu 'All Products'.

10.3.5 Submenu – Customers

Figure 10-10: Submenu Structure - Customers

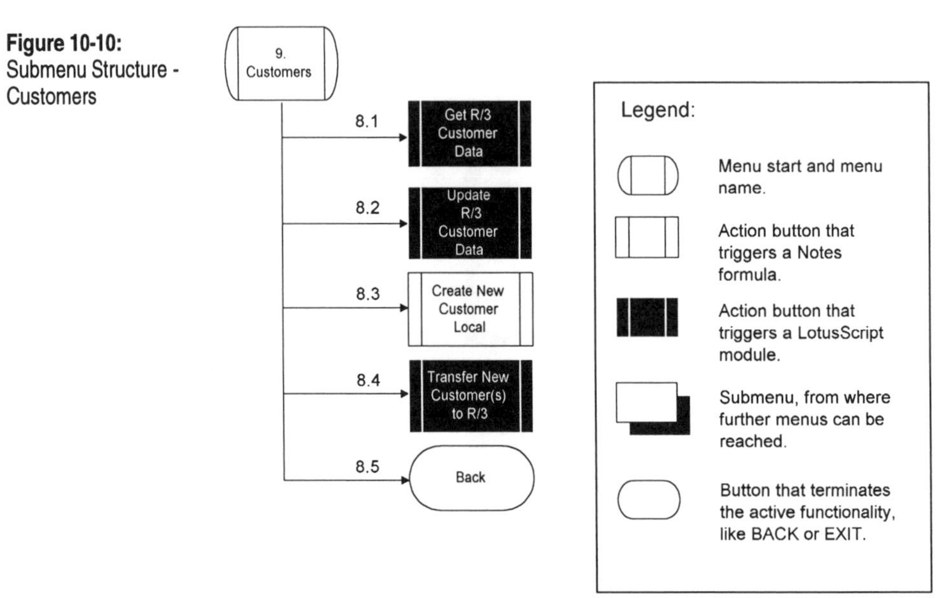

The submenu 'Customers' holds and handles copies of all R/3 customers.

The functionalites that are available in this submenu support the user with the following tasks:

- **Get R/3 Customer Data** – retrieves all customer information from the R/3 system into Lotus Notes
- **Update R/3 Customer Data** – updates changed Notes customer records in the R/3 system.
- **Create New Customer Local** – creates a new customer locally in Notes.

Transfer New Customer(s) to R/3 – all new customers can be transferred to the R/3 system.

10.3.6 Submenu – Orders

Figure 10-11: Submenu Structure - Orders

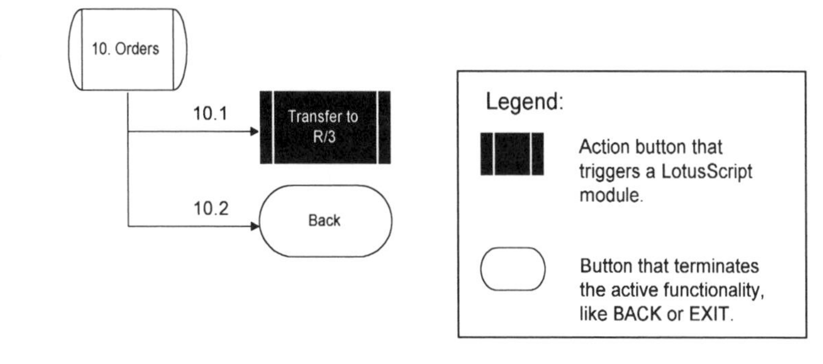

This part of the application keeps the locally created orders, before they are sent to the R/3 system to be created.

The Action Button 'Transfer to R/3' has the functionality to create customer orders on the R/3 system via the BAPI CustomerOrder.Create. The retrieved customer order number will then be inserted into the referring Notes document as a reference that the order was created successfully on the R/3 side.

10.3.7 Submenu – Setup R/3 Link

Figure 11-12: Submenu Structure – Setup

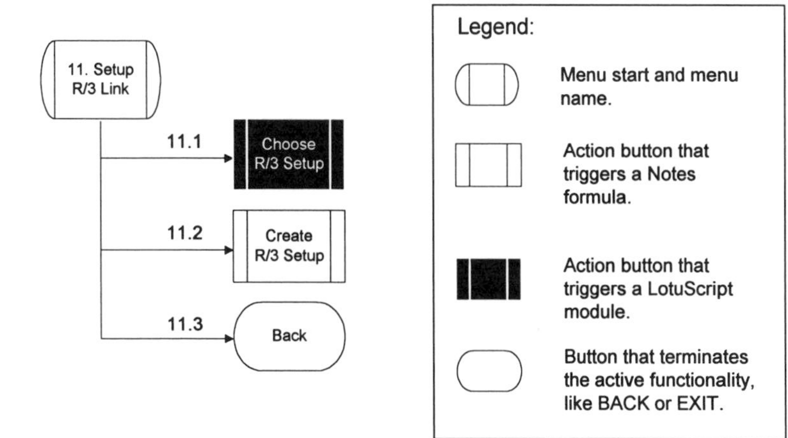

This part of the application enables the user to create or choose an already existing setup for a R/3 system. The selected R/3 setup configuration will then be stored in the notes.ini file.

11 Prototype – Technical Description

This chapter describes the technical realisation of the TES application. It describes how the *'LSX for SAP R/3'*[42] works, how certain functionality has to be programmed and how the R/3 functions (RFCs, BAPIs and Transactions) have to be called to exchange data between a Lotus Notes Client and a R/3 System.

Figure 11-1: TES Technical Architecture

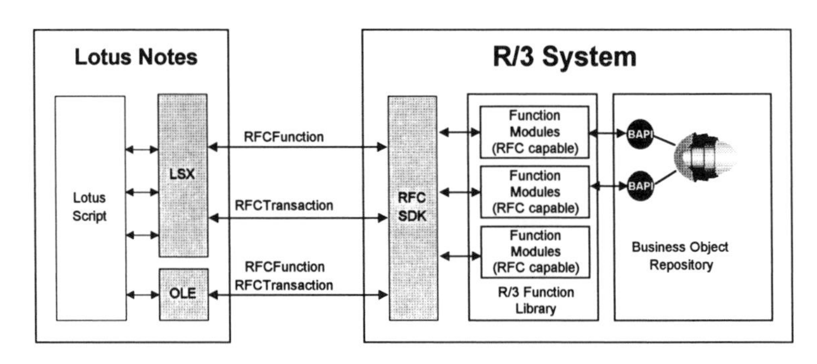

The Lotus Script Extension (LSX) for SAP R/3, that was used, enables an application to access SAP R/3 data bi-directionally by using Lotus Notes or any of the Lotus SmartSuite products so that applications can be built that read and/or update data in a SAP R/3 System.

The 'LSX for SAP R/3' is an object-oriented plug-in for Lotus Script that utilises the functionality provided by *RFCSDK* (Remote Function Call Software Development Kit) of SAP. The RFCSDK calls up the R/3 function modules in the R/3 function library, which are referenced in the function call.

Then the R/3 Function Library sends messages to the appropriate BAPI, which then triggers the associated method of the corresponding Business Object. Return values will then be delivered in the same way back from where the initiated call was sent.

It is also possible to trigger SAP-transactions through the LSX class 'RFCTransaction'. The TES system utilises this technique in the function 'Transfer New Customer(s) to R/3' which is described in 10.6

[42] 'LSX for SAP R/3' is available free of charge on the R/3 presentations CD starting from Release 3.1G

As an example of how to program a successful LSX – R/3 connection, the function module 'Get Product Details' will be taken, which was already described theoretically in section 10.3 of this book.

11.1 How to use LSX for SAP R/3

Note that in this section the code will be explained in words before the corresponding line of code.

To use the 'LSX for SAP R/3' within an application, the "USELSX" statement needs to be added to the [Options]-section, which can be selected from the *Event Box*[43] of the Design Pane in Notes' ID. The USELSX statement offers the following two options of usage:

1. Enter the LSX name as a library filename including the full path like C:\NOTES\NRFCLSX.DLL.
2. Use the name that is associated with the LSX in the LSX class registry.

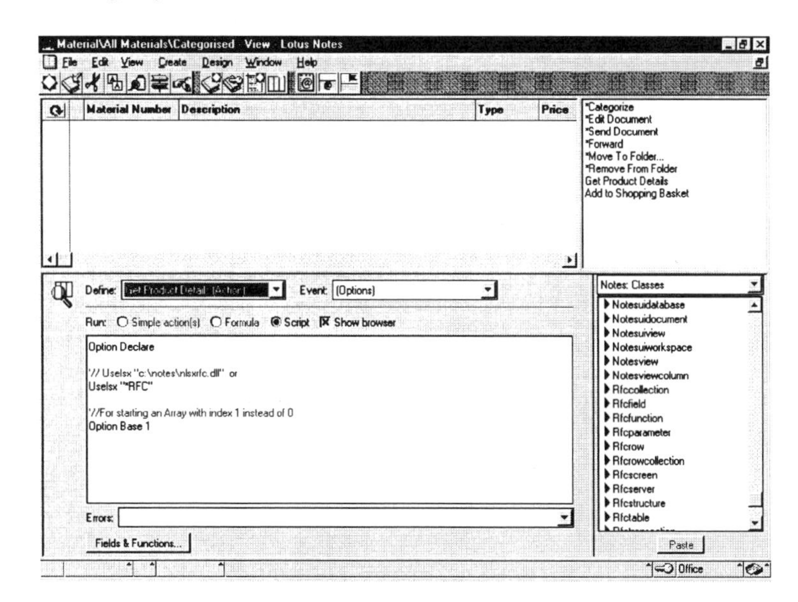

Figure 11-2: The Lotus Script Editor and Browser

[43] The Event box shows all of the events that are associated with the current object or action.

If for example the LSX is registered within the system registry as in Rfc=C:\NOTES\NRFCLSX.DLL or a similar path, the statement is as follows:

```
Uselsx "*RFC"
```

We chose the second option because it is much more location independent. On leaving the [Options]-section, the Lotus application loads and registers the LSX. The LSX classes that is to say RFC-Classes will then be represented in the IDE class browser, which is shown in the figure 11-2.

11.1.1 Function Modules used to access R/3's BAPIs

As mentioned previously, the function 'Get Product Details', located in the submenu 'All Products', will be used to describe the approach to develop an application that is able to connect Lotus Notes with an R/3 system.

To meet the constraints of the page size of this report, this symbol ↳ is used to indicate a line of program code that was too long to fit on a single printed line but that must appear as one line in the code.

The coding of the core TES-LotusScript module that was described on formal level in chapter 10 looks as follows:

```
Sub Click(Source As Button)
     If OpenConnection() = True Then
          Call GetMaterial()
          Call CloseConnection()
     Else
          Msgbox ("ERROR: Cannot get a connection to
            ↳ the specified R/3 system!!")
     End If
End Sub
```

The whole functionality is triggered through the Click event, which is associated with the Action button 'Get Product Details'.

11.1.2 Function – OpenConnection

This function verifies whether an active server object has been initialised, which references an active connection to an R/3 server. If an established connection is available, this will be used. If not, a new global server object will be created and connected to an R/3 server, using the settings of the active logon setup, chosen in the view Administration/R/3 Logon Setup.

```
Function OpenConnection As Integer
```

The declaration of a variable is possible either explicitly or implicitly. Using the **Dim** statement variable, names are declared explicitly. To be sure that the used variables are declared with the right datatype, an explicit declaration is preferred.

```
Dim intRet As Integer
Dim strSAPSystem As String
Dim strSAPServer As String
Dim strSAPSysno As String
Dim strSAPClient As String
Dim strSAPUser As String
Dim strSAPLanguage As String
Dim strSAPPassword As String
```

The variable **objServer** is defined in the [Declarations]-section of this function with the result that it is globally available within all functions. The statement in the [Declarations]-section looks as follows:

```
Dim objServer As RFCServer
```

It may be possible that a connection to an R/3 system was established and is currently running. To prevent possible collisions, this check has to be made and a second connection has to be prevented.

```
If Not (objServer Is Nothing) Then
```

The following **IF** statement checks if there is a running connection and if so whether it is connected to an R/3 system. If it is connected then stop trying to connect to the R/3 system and use the current one.

```
    If objServer.Connected = True Then
      OpenConnection = True
      Exit Function
    End If
End If
```

The next statement creates a **NotesSession** object for retrieving the logon settings from the Notes environment. The following information is stored and is read from the **notes.ini** file. The following $-variables are an excerpt of the **notes.ini** file that stores the information about the selected R/3 setup.

```
$SAPSysno=00
$SAPUser=testuser
```

```
$SAPPassword=test
$SAPClient=045
$SAPLanguage=E
$SAPServer=asec-hp3
```

The problem with this approach is that the SAP password is stored in the notes.ini file as well. This is dangerous, because the notes.ini file is not protected against misuse. The LSX documentation recommends that you avoid storing the password in the notes.ini file but does not suggest a better alternative.

The NotesSession class represents the Notes environment of the current script, providing access to environment variables, address books, information about the current user, and information about the current Notes platform and release number.

```
Dim Session As New NotesSession
```

Fetching the logon settings is done using the **GetEnvironmentString** method which retrieves the value of an environment variable.

```
strSAPSystem = Session.Getenvironmentstring
   ↳ ("SAPSystem")
strSAPServer = Session.Getenvironmentstring
   ↳ ("SAPServer")
strSAPSysno = Session.Getenvironmentstring
   ↳ ("SAPSysno")
strSAPClient = Session.Getenvironmentstring
   ↳ ("SAPClient")
strSAPUser = Session.Getenvironmentstring
   ↳ ("SAPUser")
strSAPPassword = Session.Getenvironmentstring
   ↳ ("SAPPassword")
strSAPLanguage = Session.Getenvironmentstring
   ↳ ("SAPLanguage")
```

This checks if an active logon information is setup properly. If the concatenated logon string is empty there are no entries in the notes.ini file.

```
If strSAPSystem + strSAPServer + strSAPSysno
   ↳ + strSAPClient + strSAPUser
   ↳ + strSAPPassword + strSAPLanguage = "" Then
OpenConnection = False
   Msgbox "No active logon configuration.
      ↳ Please select one in the 'R/3 Logon
```

```
    ↳ Setup' menu."
      Exit Function
   End If
```

The variable **objServer** is defined in the [Declarations]-section of this function with the result that it is globally available within all functions. The following line uses the objServer object, creates an instance and applies it to the objServer object.

```
Set objServer = New RFCServer
```

The **RFCServer** object handles the properties that are necessary to address the R/3 Application Server. The above-mentioned statement creates the server object and fills it up with the necessary properties for a R/3 logon.

```
objServer.Destination = strSAPSystem
objServer.HostName    = strSAPServer
objServer.System      = Val(strSAPSysno)
objServer.Client      = strSAPClient
objServer.User        = strSAPUser
objServer.Language    = strSAPLanguage
objServer.Password    = strSAPPassword
```

After receiving all necessary logon data, a logon can be made to the selected R/3 system.

```
intRet = objServer.Logon()
If intRet <> True Then
   Msgbox objServer.Message, 0, "Logon error"
End If
   OpenConnection = intRet
End Function
```

After setting up a successful connection to a R/3 system, certain R/3 functionalities can be triggered via RFC functions (respectively BAPIs) or via SAP Transactions. Both of them are used in the TES application but the functional approach will be described first.

11.1.3 Sub[44] – GetMaterial

The Sub GetMaterial is responsible for retrieving the material details with the help of the BAPI Material.GetDetail. In other

[44] The main difference between a function and a sub within LotusScript is that a function returns a value and a sub does not.

words the BAPI Material.GetDetail is the method that provides detailed data for a specific material.

```
Sub GetMaterial
```

It declares objects for the creation of Notes documents and other Notes related variables.

```
Dim Session As New NotesSession
```

The **NotesUIWorkspace** represents the current Notes workspace window. It empowers LotusScript to perform some of the actions that are normally performed from the Notes workspace, such as opening databases or creating and editing documents.

```
Dim ws As New NotesUIWorkspace
```

The **NotesDatabase** represents a Notes database and provides properties and functionality to access (create, modify and delete) them.

```
Dim dbCurr As NotesDatabase
```

The **NotesDocument** is a core class that represents a document in a database and provides certain functionalities for accessing and modifying it.

```
Dim docCurr As NotesDocument
```

The **RFCFunction** class is used to create an RFC function. It contains everything that is needed to make an RFC call. When the function object is created, all appropriate Import/Export parameters (see RFCParameter class) as well as all appropriate table objects (see RFCTable class) will be created automatically.

```
Dim RfcMaterialGet As RFCFunction
```

The Import or Export-parameter of a RFCFunction may be either of the type RFCParameter or of the type RFCStructure. If an Import/Export is of the type RFCStructure, the structure itself depends on the RFC function it refers to. In other words, you must check which structure parameters can be used.

This declares the table object

```
Dim structMaterial As RFCStructure
Dim view As NotesView
```

Declaring Material as an array and type string

```
Dim arrMaterial () As String
Dim NumMaterial  As String
```

```
Dim intEndRow As Integer
Dim Counter As Integer
```

Initialising the Counter variable with 1.

```
Counter = 1
Set dbCurr = Session.CurrentDatabase
```

Set view to the Notes-view 'Catalog', where all R/3 material numbers are stored, which are set as import parameters for the BAPI Material.GetDetail.

```
Set view = dbCurr.GetView("Catalog")
```

Set docCurr so that it points to the first document in view.

```
Set docCurr = view.GetFirstDocument
```

Run through the While-Loop as long as the view 'Catalog' (see TES submenu Material Catalog) contains documents. This Loop fills up the array 'arrMaterial' with all material numbers stored in the 'Catalog' view, which will be supported as import parameters to the BAPI later on.

```
While Not (docCurr Is Nothing)
```

Redim, that stands for **re-dim**ension the array 'arrMaterial' to the value stored in the 'Counter' which will then be the array's upper bound. The **Preserve** command is responsible for keeping the existing entries.

```
    Redim Preserve arrMaterial(1 To Counter)
        arrMaterial(Counter) =
          ↳ docCurr.Columnvalues(0)
```

Set docCurr to the next document in the view 'Catalog'.

```
        Set docCurr =
          ↳  view.GetNextDocument(docCurr)
```

Increment the 'Counter' variable by 1.

```
        Counter = Counter + 1
Wend
```

Get the size of arrMaterial and get number of rows to be imported.

```
intEndRow = Ubound(arrMaterial)
```

Create the function object and bind it to the R/3 function module BAPI Material.GetDetail. The first parameter objServer is a global variable that was previously set by the function OpenConnection. It specifies the R/3 application server on which the function module will be processed.

```
Set RfcMaterialGet =
    ↳ New RFCFunction(objServer,
    ↳ "BAPI_MATERIAL_GET_DETAIL")
```

Loop through arrMaterial, retrieve data and create a Notes document for each row.

```
For Counter = 1 To intEndRow Step 1
```

Assign the Import-parameters of the function module for data selection. To understand this handling of parameters a little bit better, the whole technique has to be seen from the client's side: The server-function-module' s (R/3 side) Import-parameters are the Export-parameters (TES side) and vice versa. Figure 11-3 will help to illustrate this statement.

Figure 11-3: Client and Server Import / Export Parameters

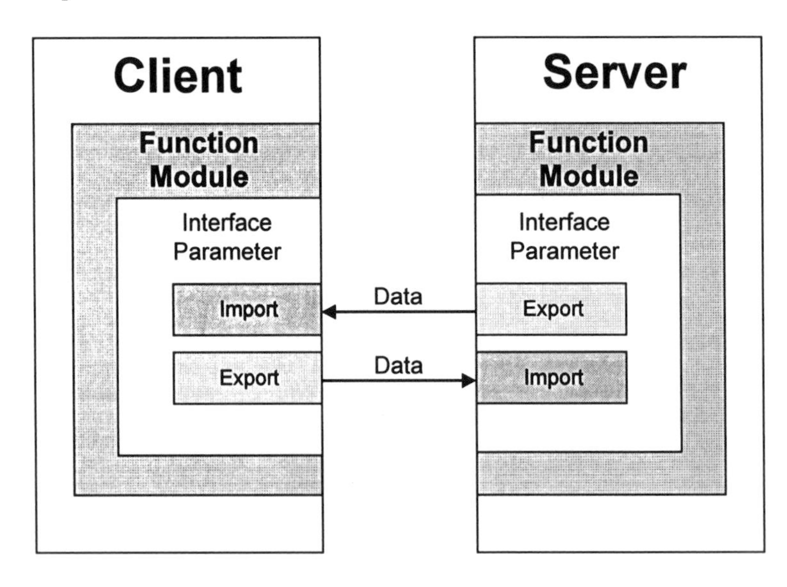


```
RfcMaterialGet.Exports("MATERIAL").Value =
    ↳ arrMaterial(Counter)
Print "Retrieving Material data, please
    ↳ wait..."
```

The Import/Export-parameters from the BAPI Material.GetDetail can be seen in figure 11-4.

Call the BAPI and if the result is true, process the data, if not pop up a message.

```
If RfcMaterialGet.Call = True Then
```

Binding the structure-object to the BAPIs Export interface that is called MATERIAL_GENERAL_DATA.

```
Set structMaterial =
  ↳ RfcMaterialGet.MATERIAL_GENERAL_DATA
Set docCurr = dbCurr.CreateDocument
```

Set various document fields that define which form is used to create a new document in the database where the retrieved BAPI information will be stored.

```
docCurr.Type = "MATERIAL"
docCurr.Form = "MATERIAL"
```

Assignment of the retrieved data to the corresponding document fields can be done in two ways:

Example (1):

```
docCurr.fldMaterialNum =
↳ RfcMaterialGet.Exports("MATERIAL").Value
```

Example (2):

```
docCurr.MAT_DESC = RfcMaterialGet.
↳ MATERIAL_GENERAL_DATA.
↳ GetValue("MATL_DESC")
```

We will stick to example (1) because it is shorter and also easier to read and to understand.

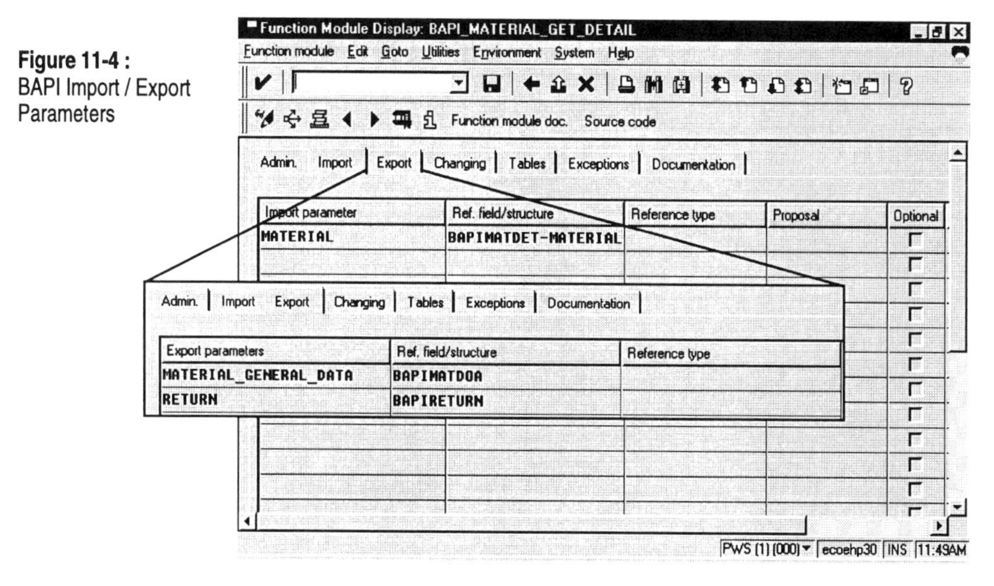

Figure 11-4 : BAPI Import / Export Parameters

```
docCurr.MAT_TYPE =
  ↳ structMaterial.GetValue("MATL_TYPE")
 docCurr.MAT_GROUP =
  ↳ structMaterial.GetValue("MATL_GROUP")
 docCurr.BASE_UOM =
  ↳ structMaterial.GetValue("BASE_UOM")
```

Note: The field OLD_MAT_NO, within the R/3 material master contains the Material Price. This trick was necessary because the BAPI Material.GetDetail did not provide the sales-price for a certain material.

```
docCurr.PRICE =
 ↳ structMaterial.GetValue("OLD_MAT_NO")
docCurr.GROSS_WT =
 ↳ structMaterial.GetValue("GROSS_WT")
docCurr.NET_WEIGHT =
 ↳ structMaterial.GetValue("NET_WEIGHT")
docCurr.UNIT_OF_WT =
 ↳ structMaterial.GetValue("UNIT_OF_WT")
docCurr.Categories =
 ↳ structMaterial.GetValue("PROD_HIER")
```

Saving the new document in the database.

```
  If Not (docCurr.Save( True, True ) = True)
   Then
     Messagebox("Document  was not saved")
 End If
```

A Refreshing of the Notes-view 'Material' is carried out by the following command.

```
 Call ws.ViewRefresh()
Else
```

In the case of an error the "Message" property contains the appropriate error message.

```
   Msgbox "Call Failed! error: " +
     ↳ RfcMaterialGet.Message
  End If
 Next
End Sub
```

11.1.4 Sub – CloseConnection

This function closes the synchronous connection from the TES-Client to the R/3 Server

```
Sub CloseConnection()
    Call objServer.Logoff
End Sub
```

with the help of the RFCServer-method **Logoff**, which disconnects the TES-System from the SAP-System.

In the functionality, described above, the question of how to access BAPIs or RFC-functions on a R/3 server was one problem, which was encountered. The problem is that an insufficient number of BAPIs are available at the moment (in Release 3.1 G) to cover the needs of this TES application. This was especially encountered in the section where a customer can be created in the TES system locally and no BAPI was available in the R/3 system. In other words, this BAPI Create.Customer is not existent at the moment. A solution to this problem was to call up directly from the SAP-transaction 'VD01' that creates a customer in the R/3 system. This technique of calling up and using SAP-transactions directly will be described in the following section.

11.1.5 R/3 Transactions – How can they be accessed?

It is assumed that the connection between the TES application and the R/3 system has been successfully established and can be used to call the following SAP-transaction 'VD01' that allows the user to create a customer on the R/3 system. This interfacing technique is also known as 'Call Transaction'.

Call Transaction

Before this kind of functionality can be used, it is necessary to identify, which SAP transaction screens have to be called up to create a customer successfully on the R/3 system.

The R/3 system provides the user with a so-called "Batch Input Recording", which has to be started before a SAP transaction. This records the screen sequence and the data entries of the user during a transaction. Figure 11-5 shows an excerpt of the recorded transaction entries, which can be taken, as a source for realising the screen sequence of a SAP transaction, within LotusScript.

Note that some of the code sections are itemised more than once. The reason for that is that the reader should get a feeling for the whole complexity, and what has to be considered when creating proper master data via SAP-transactions.

```
Sub TACreateR3Customer
  Const CTransCode = "VD01"
  Dim workspace As New NotesUIWorkspace
  Dim Session As New NotesSession
  Dim dbCurr As NotesDatabase
  Dim viewCust As NotesView
  Dim docCurr As NotesDocument
  Dim FieldValue(0)  As String
  Dim positionOfChar As Integer
```

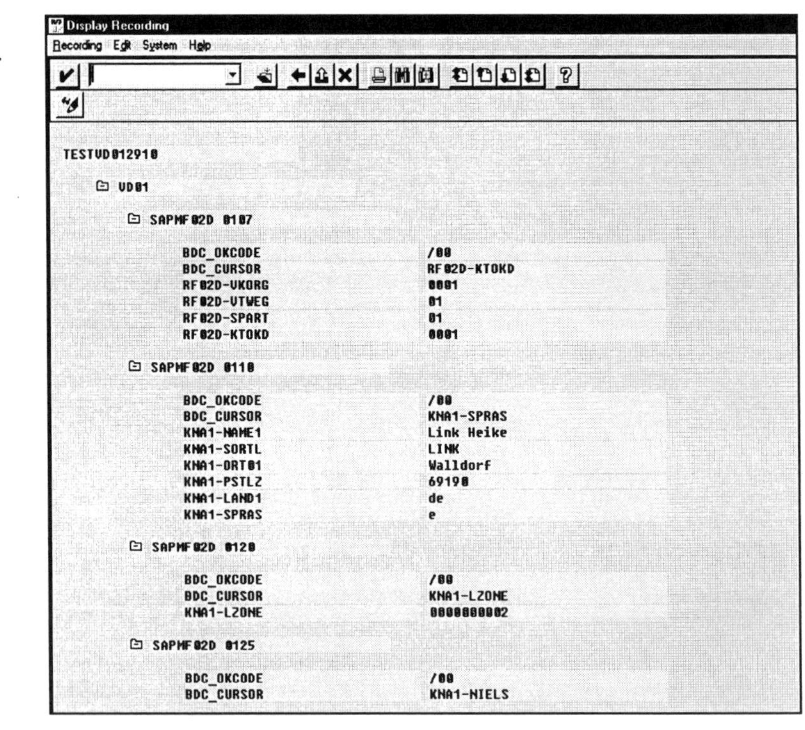

Figure 11-5:
The 'Batch Input Recording' result from transaction VD01

The **RFCTransaction** class supports batch mode transaction processing on a R/3 system. A RFCTransaction object handles one R/3-Transaction and can contain multiple **RFCScreen** objects, which represents the different SAPGUI Dynpros also known as screens, belonging to a SAP-transaction.

```
Dim TransAction As New RFCTransaction
  ↳ (objServer, CTransCode)
```

The RFCTransaction, mentioned above, uses the RFCScreen object. On creation of the RFCTransaction object it will be initialised with all screens that refer to the given transaction.

```
Dim Screen As RFCScreen
Set dbCurr = Session.CurrentDatabase
Set viewCust = dbCurr.GetView("Customers")
Set docCurr = viewCust.GetFirstDocument()
```

This function takes all newly created customers and creates them in the R/3 System.

```
While Not ( docCurr Is Nothing )
 If docCurr.IsNew(0) = "1" Then
  Print "Creating customer " +
    ↳ docCurr.NAME1(0) + "in the R/3
    ↳ System..."
```

The following coding shows how a transaction screen is called up and activated using the enabled property of the RFCScreen class.

```
Set Screen = TransAction.Screens("0107")
Screen.Enabled = True
```

The different fields then have to be filled, including at least the mandatory fields on a screen, and then have to be confirmed with the **BDC_OKCODE** field. The BDC_OKCODE is a special field for representing entries like 'enter' or 'save' in the dialog mode of the transaction. The BDC_OKCODE "/00" represents the command 'next screen' and "/11" to stop and save the transaction. All the customer information will be stored in the R/3 database table KNA1.

```
Screen.Fields("RF02D-VKORG").Value = "0001"
Screen.Fields("RF02D-VTWEG").Value = "01"
Screen.Fields("RF02D-SPART").Value = "01"
Screen.Fields("RF02D-KTOKD").Value = "0001"
Screen.Fields("BDC_OKCODE").Value = "/00"
Set Screen =TransAction.Screens("0110")
Screen.Enabled = True
Screen.Fields("KNA1-ANRED").Value =
  ↳ docCurr.ANRED(0)
Screen.Fields("KNA1-NAME1").Value =
  ↳ docCurr.NAME1(0)
```

In this section of the code the last and first names of a customer are extracted from the Name-field.

```
positionOfChar = Instr(1, docCurr.NAME1(0), " ")
If (positionOfChar = 0) Then
  Screen.Fields("KNA1-SORTL").Value =
   ↳ docCurr.NAME1(0)
Elseif (positionOfChar <= 10) Then
  Screen.Fields("KNA1-SORTL").Value =
   ↳ Left$(docCurr.NAME1(0),
         ↳ positionOfChar)
Else
```

If a customer has a last name longer then 10 characters, the first 10 of them will be extracted and put into the screen field 'SEARCH TERM' (KNA-SORTL).

```
  positionOfChar = 10
  Screen.Fields("KNA1-SORTL").Value =
    ↳ Left$(docCurr.NAME1(0), positionOfChar)
End If
  Screen.Fields("KNA1-STRAS").Value =
    ↳ docCurr.STRAS(0)
```

Check if the field 'KNA1-PFACH' has an entry, then the field 'KNA1-PSTL2' also has to be filled.

```
If (Len(docCurr.PFACH(0)) > 0) Then
  Screen.Fields("KNA1-PFACH").Value =
    ↳ docCurr.PFACH(0)
  Screen.Fields("KNA1-PSTL2").Value =
    ↳ docCurr.PFACH_ZIP(0)
End If
  Screen.Fields("KNA1-ORT01").Value =
    ↳ docCurr.ORT01(0)
  Screen.Fields("KNA1-PSTLZ").Value =
    ↳ docCurr.PSTLZ(0)
```

Country and Language is set to the TES applications default values DE (Germany) and E (English).

```
Screen.Fields("KNA1-LAND1").Value = "DE"
Screen.Fields("KNA1-SPRAS").Value = "E"
Screen.Fields("KNA1-TELF1").Value =
  ↳ docCurr.TELF1(0)
Screen.Fields("KNA1-TELFX").Value =
  ↳ docCurr.TELFX(0)
Screen.Fields("BDC_OKCODE").Value = "/00"

Set Screen = TransAction.Screens("0120")
```

```
Screen.Enabled = True
```

The mandatory field transport zone is set to the 'south region'.

```
Screen.Fields("KNA1-LZONE").Value = "0000000002"
Screen.Fields("BDC_OKCODE").Value = "/00"
```

The following 5 screens do not need any entries so that you can skip these using the 'next screen' command.

```
Set Screen = TransAction.Screens("0125")
Screen.Enabled = True
Screen.Fields("BDC_OKCODE").Value = "/00"

Set Screen = TransAction.Screens("0340")
Screen.Enabled = True
Screen.Fields("BDC_OKCODE").Value = "/00"

Set Screen = TransAction.Screens("0370")
Screen.Enabled = True
Screen.Fields("BDC_OKCODE").Value = "/00"

Set Screen = TransAction.Screens("0360")
Screen.Enabled = True
Screen.Fields("BDC_OKCODE").Value = "/00"

Set Screen = TransAction.Screens("0310")
Screen.Enabled = True
Screen.Fields("BDC_OKCODE").Value = "/00"

Set Screen = TransAction.Screens("0315")
Screen.Enabled = True
```

The R/3-field 'Shipping Condition' is set to standard, as represented in the R/3 database field 'KNVV-VSBED'.

```
Screen.Fields("KNVV-VSBED").Value = "02"
Screen.Fields("BDC_OKCODE").Value = "/00"

Set Screen = TransAction.Screens("0320")
Screen.Enabled = True
Screen.Fields("BDC_OKCODE").Value = "/00"

Set Screen = TransAction.Screens("1350")
Screen.Enabled = True
```

Tax Condition (KNVI-TAXKD) is set to standard, as represent by the value 0.

```
FieldValue(0) = "0"
Screen.Fields("KNVI-TAXKD").Value = FieldValue
```

```
Screen.Fields("BDC_OKCODE").Value = "/11"
```

The only method in the RFCTransaction class is the Call method. It executes the transaction and stores the above mentioned screen fields in the R/3 system. The mode for calling the transaction can be synchronous ("S") or asynchronous ("A").

```
      If (TransAction.Call ("S") = True) Then
       docCurr.KUNNR = Mid$(TransAction.Message, 9, 10)
       docCurr.IsNew = "0"
       docCurr.IsModified ="0"
       Print "Transaction completed and customer "
             ↪ Mid$(TransAction.Message, 9, 10)
             ↪ " in R/3 created !"
       Call docCurr.Save(True, True)
       Call workspace.ViewRefresh
      Else
       Print "ERROR: Transaction wasn't successful!"
      End If
    End If
    Set docCurr = viewCust.GetNextDocument( docCurr )
   Wend
  End Sub
```

11.2 Connections to a R/3 System via OLE

Even if the connectivity between Lotus Notes and SAP is based on the LSX, there is also a way around it. In other words the OLE classes, which are provided by SAP with the installation of their SAPGUI, can also be used to build the functionality between these two systems. The following coding example shows how that can be done:

```
Call tableOrderPartners.SetCell(Counter,"PARTN_NUMB",
    ↪ "0000000032")
tempStr = tableOrderPartners.GetCell
    ↪ (Counter,"PARTN_NUMB")
```

The goal was to use the OLE classes instead of the LSX. The following coding comparison, which is only done with the important coding, should show the difference between the two approaches.

The example that is used to compare these two approaches is the code used behind the Action button ‘Transfer to R/3’.

The following tables are taken as an example to compare the LSX approach against the OLE-approach of the functionality that opens a connection between Lotus Notes and the assigned R/3 System. The name for the function is called, in the 'active coding', '*OpenRFCConnection*'.

<table>
<tr><td>LSX</td><td>Dim objServer As RFCServer</td></tr>
<tr><td>OLE</td><td>Dim conObject As Variant</td></tr>
<tr><td colspan="2">This coding visualises the different settings in the [Declarations]-section of the two approaches. The LSX provides a datatype that handles RFCServer settings, whereby the OLE-classes use a Variant[45] datatype to store this information.</td></tr>
</table>

<table>
<tr><td>LSX</td><td>Set objServer As New RFCServer</td></tr>
<tr><td>OLE</td><td>Set oSAPLogonControl = CreateObject
↳ ("SAP.LogonControl.1")
Set conObject = oSAPLogonContorl.NewConnection</td></tr>
<tr><td colspan="2">The necessary variable setting for the instantiation of a Notes-to-R/3 connectivity is done. Within the OLE-coding you have to create an object for the SAP-Logon whereas this can be done in one step through the instantiation of the RFC Server-class in the LSX.</td></tr>
</table>

<table>
<tr><td>LSX</td><td>intRet = objServer.Logon()
If intRet <> True Then
...
End If
OpenConnection = intRet</td></tr>
<tr><td>OLE</td><td>If conObject.Logon(0, True) = False Then
... '// R/3 connection failed
Else
... '// R/3 was succesful
End If</td></tr>
<tr><td colspan="2">These two codings for checking whether an already existing Notes-to-R/3 connection is active or not. In case one active connection is available this one will be taken in the other case a new connection will be opened.</td></tr>
</table>

[45] The datatype Variant specifies a 16-byte variable that can contain data of any scalar type, an array, a list, or an object.

The following coding examples are taken from the function that passes the order parameter on to the BAPI CustomerOrder.Create. The name for the function is, in an OLE-solution, '*OrderCreate*'.

LSX

```
Dim RfcOrderCreate As RFCFunction
Set RfcOrderCreate = New
  ↪ RFCFunction(objServer,
  ↪ "BAPI_CUSTOMERORDER_CREATE")
```

OLE

```
Dim oOrder As Variant
Dim oSAPFunction As Variant
Set oSAPFunction =
 ↪ CreateObject("SAP.Functions")
Set oSAPFunction.Connection =
 ↪ conObject
Set oOrder = oSAPFunction.Add
 ↪ ("BAPI_CUSTOMERORDER_CREATE")
```

Here are two examples for the declaration and initialisation of an RFC- respectively SAP-Function to the BAPI CustomerOrder.Create.

LSX

```
Dim structOrderHeaderIn As
 ↪ RFCStructure
Set structOrderHeaderIn =
 ↪ RfcOrderCreate.Exports
  ↪ ("ORDER_HEADER_IN")
structOrderHeaderIn = cDOC_TYPE
```

OLE

```
oOrder.Exports
 ↪ ("ORDER_HEADER_IN").Value
  ↪ ("DOC_TYPE") = cDOC_TYPE
```

This is the point, where one of the export parameters from the above-mentioned BAPI is filled.

LSX

```
Dim structReturn As RFCStructure
Set structReturn = RfcOrderCreate.RETURN
If structReturn.GetValue("TYPE") = "" Then
… '// Order created successfully
Else
… '// Order wasn't successfully created
End If
```

OLE

```
If oOrder.Import("RETURN").Value
  ↳ ("TYPE") = "" Then
… '// Order created successfully
Else
… '// Order wasn't successfully
      ↳ created
End If
```

This example shows how the export values from the above-mentioned BAPI, in the "RETURN"-structure, are retrieved and checked.

To conclude this section of the chapter it can be said that when the LSX – and the OLE – approach are compared, that these two ways of accessing R/3 functionality are almost the same. In other words, the effort needed to program the functionality in both cases is actually the same.

11.3 Customized RFC Functions

This section describes how to program a RFC-Function within the R/3 system, when a certain functionality is not provided by the SAP. The synchronous RFC-technology has already been described within chapter 9 of this documentation, so that a more detailed description of this technique is not necessary.

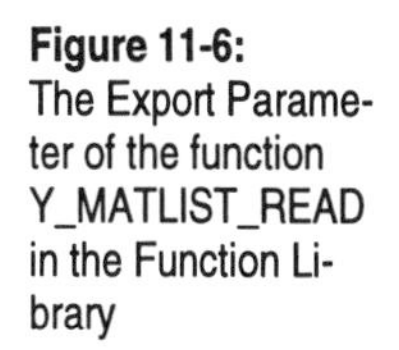

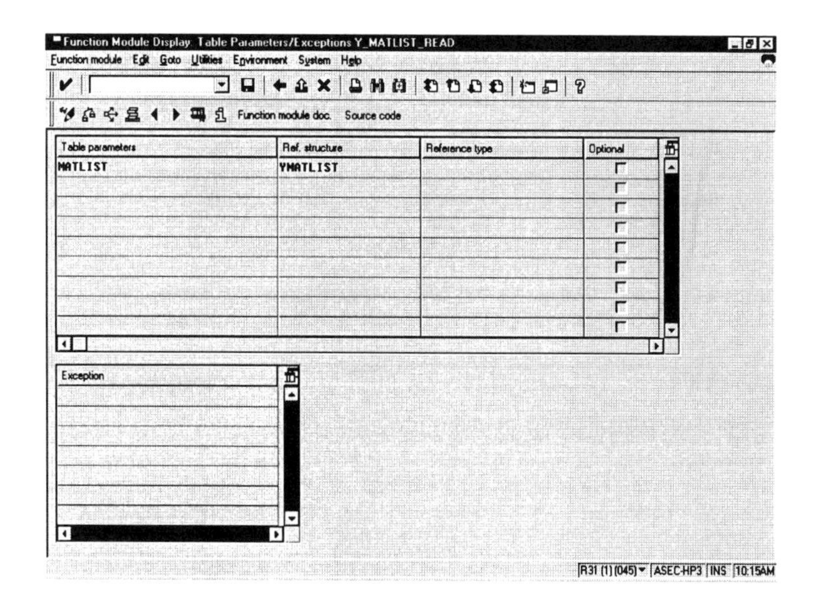

Figure 11-6: The Export Parameter of the function Y_MATLIST_READ in the Function Library

This function is used by the Action button 'Get R/3 Material Numbers' to retrieve all material numbers that start with the letters 'GM'. This is necessary because the import parameter of the BAPI 'Material.GetDetail' is a product/material number. So the user has two ways of providing this material number to the R/3 system, either he/she keys in every single number or the necessary functionality e.g. as follows, is programmed.

11.3.1 RFC function Y_MATLIST_READ

This ABAP/4 function represents how easily own functions can be included and accessed with the RFC functionality. RFC_GET_TABLE_ENTRIES and RFC_READ_TABLE would supply a similar kind of functionality but as mentioned, this function is an example of how an ABAP/4 function can be accessed.

```
function y_matlist_read.
 *"----------------------------------------------------
 *" Author   : Gerd Moser
 *" Version  : 1.0
 *" Function : Retrieves all material numbers
 *" starting with GM and writes them into the table
 *" 'matlist'
 *"----------------------------------------------------
 *" Local Interface
 *"    TABLES
 *"       MATLIST STRUCTURE  YMATLIST
*"-----------------------------------------------------
```

The following SQL-statement retrieves, as mentioned above, all products starting with the two letters 'GM' from the R/3-table **'mara'**. This table is the master table that stores all main material data of an R/3 system.

```
tables: mara.
select distinct * from mara where matnr like 'GM%'.
```

The return value sy-subrc contains a 0 if the selected statement was successful, which means it returned some values, or it contains a value that is not equal to 0, if it was not successful.

```
if sy-subrc = 0.
```

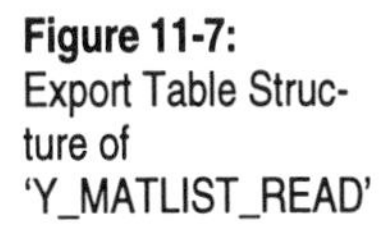

Figure 11-7: Export Table Structure of 'Y_MATLIST_READ'

The next two steps are responsible for assigning the retrieved value mara-matnr to the export table value matlist-matnr and appends them to the table matlist, which is shown in figure 11-6. The structure of the table matlist can be seen in figure 11-7.

```
      matlist-matnr = mara-matnr.
         append matlist.
       endif.
     endselect.
  endfunction.
```

Before the function can be accessed through LotusScript Y_MATLIST_READ has to be defined as a RFC enabled function.

This is done using the button 'Remote Function Call Supported' in the administration area of the function module's definition (Transaction SE37) as shown in the figure.

11.3.2 Sub Get_R3_MaterialNumbers

This Sub is now the pendant to the RFC function module 'Y_MATLIST_READ' and is responsible for retrieving the material numbers which are delivered.

```
Sub Get_R3_MaterialNumbers()
```

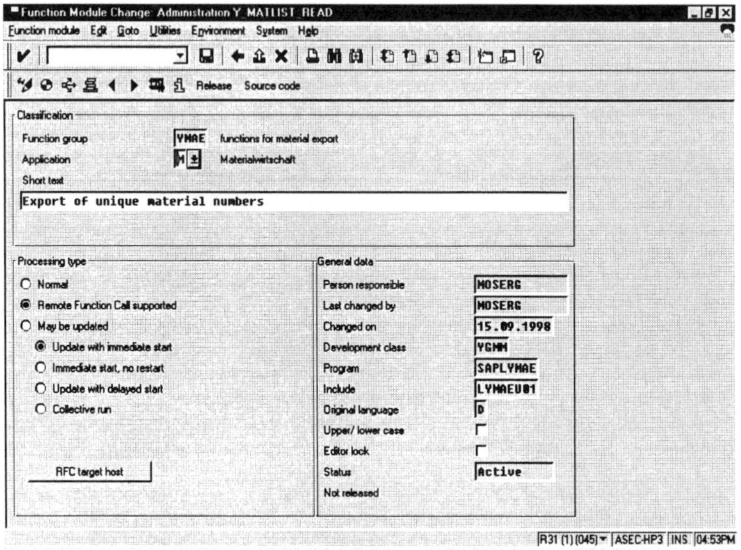

Figure 11-8: RFC enabled Function Module

It declares the objects for the creation of Notes documents and other Notes related items. NotesSession represents the Notes environment of the current script, providing an access to environment variables, address books, information about the current user, and information about the current Notes platform and release number.

```
Dim Session As New NotesSession
Dim ws As New NotesUIWorkspace
Dim dbCurr As NotesDatabase
Dim docCurr As NotesDocument
```

The **RFCFunction** class is used to create an RFC function. When the function object is created, all appropriate Import/Export parameters (see RfcParameter class) as well as all appropriate table objects (see RfcTable class) will be created automatically.

```
Dim RfcMaterialNumberGet As RFCFunction
```

The **RFCTable** object contains columns and rows. Depending on its functional context, a table may serve as input or output for an RFC function. Whenever a new function object is created, all tables that are required by that function will be created automatically.

```
Dim tblMaterialNumbers As RFCTable
Set dbCurr = Session.CurrentDatabase
```

Create the function object and bind it to the R/3 function module "Y_MATLIST_READ". The first parameter "objServer" is a global variable that is previously set by the function "OpenConnection". It specifies the R/3 application server on which the function module will be processed.

```
    Set RfcMaterialNumberGet =
      ↳ New RFCfunction(objServer, "Y_MATLIST_READ")
```

Call the function. If the result is true, process the data, otherwise pop up a message.

```
    If RfcMaterialNumberGet.Call = True Then
```

Bind the table object to the RFCs Export interface "MATLIST"

```
        Set tblMaterialNumbers =
           ↳ RfcMaterialNumberGet.MATLIST
        Set docCurr = dbCurr.CreateDocument
```

Set the document types, in which the retrieved information will be stored.

```
        docCurr.Type = "CATALOG"
        docCurr.Form = "CATALOG"
```

Loop for processing all retrieved data.

```
        Forall row In tblMaterialNumbers.Rows
          Set docCurr = dbCurr.CreateDocument
```

Set the document types, in which the retrieved information will be stored.

```
          docCurr.Type = "CATALOG"
          docCurr.Form = "CATALOG"
```

Assign the retrieved material number to the form field 'fldCatalogMatlNum'.

```
          docCurr.fldCatalogMatlNum = row.MATNR
          If Not ( docCurr.Save( True, True ) = True )
             ↳ Then
                Messagebox ( "Document  was not saved" )
          End If
```

The following command refreshes the view : 'Material'

```
          Call ws.ViewRefresh()
        End Forall
    Else
```

In the case of an error the "Message" property contains the appropriate error message.

```
        Msgbox "Call Failed! error: " +
           ↳ RfcMaterialNumberGet.Message
      End If
    End Sub
```

11.3.3 Embedding Product Images

The BAPI Material.GetDetail, provides all material details but that is not enough for this particular prototype scenario. The user should also be able to see an image of the chosen product (see figure 11-9).

Figure 11-9: Product Overview

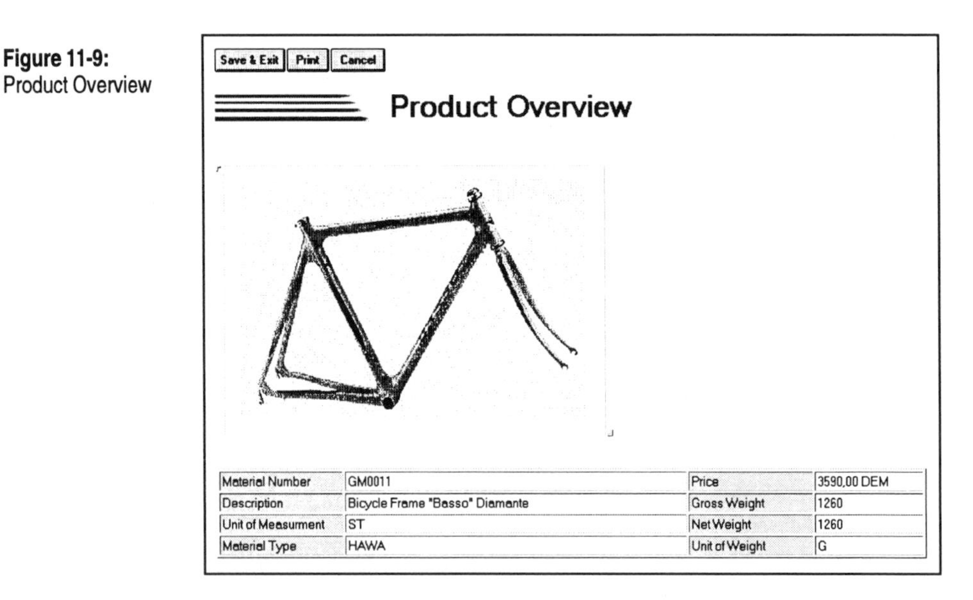

The following section describes how this idea has been realised using the OLE technology for linking a product image to its detailed data.

This function is triggered through the document's **Postopen** event.

```
    Sub Postopen(Source As Notesuidocument)
```

The next statements define the constants that are used to identify the location of the image files.

```
Const cPath   = "D:\notes\Diplomarbeit\Products\"
Const cPicExt = ".gif"
Const cMatNum = "fldMaterialNum"
```

The next statements define the constants that are used to identify the location of the image files.

```
Dim MaterialNumber As String
Dim ImagePath As String
Dim PictureFlag As String
```

Here the document field 'PictureFlag' is checked. If it contains a '0', the image is not included and has to be loaded and if it contains a '1' the following steps can be omitted.

```
PictureFlag = Source.FieldGetText("PictureFlag")
If (PictureFlag = "0") Then
  Source.EditMode = True
  MaterialNumber =
     ↳ Source.FieldGetText("fldMaterialNum")
  Call Source.GotoField( "fldMaterialPicture" )
  ImagePath = cPath + MaterialNumber + cPicExt
```

This command creates the OLE object in this document. The OLE Server for this functionality is the program Paint Shop Pro 4.1.

```
  Call Source.CreateObject("", "",  ImagePath)
```

The field 'PictureFlag' is set to '1', to avoid the product image being re-embedded when the user opens this document again.

```
  Call Source.FieldSetText("PictureFlag","1")
  Call Source.Save
Else
  Exit Sub
End If
End Sub
```

There are certain functions, which will not be described in this chapter because, as it was mentioned before, this chapter only concerns the functionality, which has to be programmed for a Lotus Notes and R/3 connection. One exception was made within this section, where the OLE functionality to embed product images was described. But it has to be mentioned that the TES prototype contains a high amount of programming effort, due to the complexity of the TES functionality that cannot be neglected.

12 Conclusion

This chapter is intended to share the experiences and lessons learned throughout this project. In addition it will also give an outlook about how the technology development of the BAPIs will continue and the TES-prototype could be further enhanced.

12.1 Lessons Learned

During the different phases of the project and whilst working with Lotus Notes, it became clear that the development of applications can be done very quickly because Lotus Notes offers quite simple design elements. To be more precise Notes supports the application developer in *Rapid Application Development* (RAD). For instance when the documents have been established, the data can be presented in different ways using the powerful views. But on the other hand, to do things that were not supported directly by Notes, the effort needed for development done in LotusScript was much greater.

An example of this is the order form of the TES prototype. When I wanted to implement a spreadsheet table into Lotus Notes, I only got a graphical representation without any additional functionality. As a result I had to develop a calculation of the sub amounts and the total amounts myself. Moreover, I had to simulate an order detail table, which included the navigation through it, the inserting of data into table rows and as already mentioned the calculation itself.

This does not mean, though, that Lotus Notes is lacking in providing support to the developer. On the contrary, one of its main advantages from which I have profited is that I did not have to be concerned about whether the application runs in an attached or detached mode from the network; Lotus Notes takes care of this complex task. Additionally I want to emphasise the Notes' capability to convert already existing applications into Internet applications on the fly via its domino server.

With respect to accessing the BAPIs from Notes using the LSX provided, it can be observed that when comparing the LSX against other development environments not all to date technical possibilities have been realised.

12.2 Future Perspectives

The success of the BAPIs as the future interfacing technology within the R/3 environment can already be seen by the great effort 3rd party vendors put into their development tools and applications to access the BAPIs. There are Lotus' LSX, Borland's Delphi, Microsoft's Visual Basic and Siebel's[46] Sales Enterprise tool to name a few.

Based on my experience, I can recommend the usage of the BAPI technology. The reader should be aware, however, that there will be changes in the naming conventions from Release 3.1G to 3.1H, but they will be stable, or even 'frozen' from Release 3.1H on. Starting with Release 4.0, BAPIs will be used as the message types within ALE. As a consequence they will be no longer restricted with respect to their connectivity (i.e. synchronous or asynchronous). SAP, however, will not manage until Release 4.x to implement all possible BAPIs.

Regarding the TES prototype it could be further enhanced by extending the customer data, which is at the moment restricted to the address of a customer. An example would be a tracking system that could store the "credit willingness", supply status of information brochures, order status including warranty, identifying the key players of a customer (i.e. CEO, CFO, etc.). In other words to implement a Sales Information System that supports the whole Sales staff with the key information needed.

Summarising there are two major benefits of the prototype. It could be used as a powerful solution for a mobile sales force as their key sales support system. Additionally, utilising Notes' WWW-functionality from its domino server allows to put the existing application into the Internet, a company's Intranet or Extranet, to enable customers to order products. This application could then be an alternative to SAP's Internet Transaction Server (ITS) or other solutions. In contrast to these the Notes solution has the advantage that it is possible to utilise all Lotus Notes functionality over the Web and as a separate database to make queries and reports independently from R/3.

In general, most R/3 systems cannot exist in isolation and therefore need some sort of interfaces to the 'outside' world. As observed on client projects many companies have no interfacing

---

[46] Siebel is one of the world's leading provider of enterprise-class sales, marketing and customer service systems [@Siebel]

strategy in place. This approach is very short-sighted with regards to their future system infrastructure as well as to the maintenance effort and costs. Therefore it is extremely important for enterprises operating globally to implement an advanced Corporate Interfacing Strategy.

I hope that this book "SAP R/3 Interfacing using BAPIs" has provided a valuable basis for understanding SAP's Interfacing Strategy and Technology.

Appendix A: BAPI Catalogue

The following BAPI Catalogue [@SAP] provides an overview of the BAPIs available to date. The catalogue lists R/3's Business Objects and their related BAPIs in alphabetical order.

Business Object: AcctngActivityAlloc

Documentation: The business object AcctngActivityAlloc is an object responsible for processes at Accounting level. It represents activity allocations performed in Controlling including all dependent postings.
Application Area: Cross-Application Components

BAPI: AcctngActivityAlloc.Check

Documentation: This method enables the user to check if an Activity Allocation can be posted in Accounting.

Availability: R/3 4.0A

BAPI: AcctngActivityAlloc.Post

Documentation: This method enables the user to post an Activity Allocation in Accounting.

Availability: R/3 4.0A

Business Object: AcctngBilling

Documentation: The business object AcctngBilling is an object responsible for processes at Accounting level. It represents the accounting-relevant part of billings posted in Sales and Distribution.

Application Area: Cross-Application Components

BAPI: AcctngBilling.Check

Documentation: This method enables the user to check if a billing can be posted in Accounting.

Availability: R/3 4.0A

BAPI: AcctngBilling.Post

Documentation: This method enables the user to post a billing in Accounting.

Availability: R/3 4.0A

Business Object: AcctngEmplyeeExpnses

Documentation: The business object AcctngEmplyeeExpnses is an object responsible for processes at Accounting level. It represents the accounting-relevant part of postings containing employee expenses in Human Resources.

Application Area: Cross-Application Components

BAPI: AcctngEmplyeeExpnses.Check

Documentation: This method enables the user to check if an employee expense can be posted in Accounting.

Availability: R/3 4.0A

BAPI: AcctngEmplyeeExpnses.Post

Documentation: This method enables the user to post an employee expense in Accounting.

Availability: R/3 4.0A

Business Object: AcctngEmplyeePaybles

Documentation: The business object AcctngEmplyeePaybles is an object responsible for processes at Accounting level. It represents the accounting-relevant part of postings containing employee payables in Human Resources.

Application Area: Cross-Application Components

BAPI: AcctngEmplyeePaybles.Check

Documentation: This method enables the user to check if an employee payable can be posted in Accounting.

Availability: R/3 4.0A

BAPI: AcctngEmplyeePaybles.Post

Documentation: This method enables the user to post an employee payable in Accounting.

Availability: R/3 4.0A

Business Object: AcctngEmplyeeRcvbles

Documentation: The business object AcctngEmplyeeRcvbles is an object responsible for processes at Accounting level. It represents the accounting-relevant part of postings containing employee receivables in Human Resources.

Application Area: Cross-Application Components

BAPI: AcctngEmplyeeRcvbles.Check

Documentation: This method enables the user to check if an employee receivable can be posted in Accounting.

Availability: R/3 4.0A

BAPI: AcctngEmplyeeRcvbles.Post

Documentation: This method enables the user to post an employee receivable in Accounting.

Availability: R/3 4.0A

Business Object: AcctngGoodsMovement

Documentation: The business object AcctngGoodsMovement is an object responsible for processes at Accounting level. It represents the accounting-relevant part of goods movements posted in Logistics.

Application Area: Cross-Application Components

BAPI: AcctngGoodsMovement.Check

Documentation: This method enables the user to check if a goods movement can be posted in Accounting.

Availability: R/3 4.0A

BAPI: AcctngGoodsMovement.Post

Documentation: This method enables the user to post a goods movement in Accounting. A commitment could be updated.

Availability: R/3 4.0A

Business Object: AcctngInvoiceReceipt

Documentation: The business object AcctngInvoiceReceipt is an object responsible for processes at Accounting level. It represents the accounting-relevant part of invoice receipts posted in Logistics.

Application Area: Cross-Application Components

BAPI: AcctngInvoiceReceipt.Check

Documentation: This method enables the user to check if a invoice receipt can be posted in Accounting.

Availability: R/3 4.0A

BAPI: AcctngInvoiceReceipt.Post

Documentation: This method enables the user to post a invoice receipt in Accounting.

Availability: R/3 4.0A

Business Object: AcctngPurchaseOrder

Documentation: The business object AcctngPurchaseOrder is an object responsible for processes at Accounting level. It represents the accounting-relevant part of purchase orders posted in Logistics.

Application Area: Cross-Application Components

BAPI: AcctngPurchaseOrder.Check

Documentation: This method enables the user to check if a purchase order can be posted in Accounting.

Availability: R/3 4.0A

BAPI: AcctngPurchaseOrder.Post

Documentation: This method enables the user to post a purchase order in Accounting. The commitments are updated.

Availability: R/3 4.0A

Business Object: AcctngPurchaseReq

Documentation: The business object AcctngPurchaseReq is an object responsible for processes at Accounting level. It represents the accounting-relevant part of purchase requsitions posted in Logistics.

Application Area: Cross-Application Components

BAPI: AcctngPurchaseReq.Check

Documentation: This method enables the user to check if a purchase requisition can be posted in Accounting.

Availability: R/3 4.0A

BAPI: AcctngPurchaseReq.Post

Documentation: This method enables the user to post a purchase requsition in Accounting. The commitments are updated.

Availability: R/3 4.0A

Business Object: AcctngRepostRevenues

Documentation: The business object AcctngRevenues is an object responsible for processes at Accounting level. It represents repostings of revenues in Controlling including all dependent postings.

Application Area: Cross-Application Components

BAPI: AcctngRepostRevenues.Check

Documentation: This method enables the user to check if revenues can be reposted in Accounting.

Availability: R/3 4.0A

BAPI: AcctngRepostRevenues.Post

Documentation: This method enables the user to repost revenues in Accounting.

Availability: R/3 4.0A

Business Object: AcctngRepstPrimCosts

Documentation: Documentation currently not available

Application Area: Cross-Application Components

BAPI: AcctngRepstPrimCosts.Check

Documentation: Documentation currently not available

Availability: R/3 4.0A

BAPI: AcctngRepstPrimCosts.Post

Documentation: Documentation currently not available

Availability: R/3 4.0A

Business Object: AcctngSenderActivity

Documentation: Documentation currently not available

Application Area: Cross-Application Components

BAPI: AcctngSenderActivity.Check

Documentation: Documentation currently not available

Availability: R/3 4.0A

BAPI: AcctngSenderActivity.Post

Documentation: Documentation currently not available

Availability: R/3 4.0A

Business Object: AcctngServices

Documentation: Documentation currently not available

Application Area: Cross-Application Components

BAPI: AcctngServices.CheckAccountAssignment

Documentation: Documentation currently not available

Availability: R/3 4.0A

BAPI: AcctngServices.PreCheckPayrollAccountAssign

Documentation: Documentation currently not available

Availability: R/3 4.0A

Business Object: AcctngStatKeyFigure

Documentation: The business object AcctngStatKeyFigure is an object responsible for processes at Accounting level. It represents postings of statistical key figures in Controlling including all dependent postings.

Application Area: Cross-Application Components

BAPI: AcctngStatKeyFigure.Check

Documentation: This method enables the user to check if statistical key figures can be posted in Accounting.

Availability: R/3 4.0A

BAPI: AcctngStatKeyFigure.Post

Documentation: This method enables the user to post statistical key figures in Accounting.

Availability: R/3 4.0A

Business Object: ActivityType

Documentation: The business object Activity type contains the dividing up (classification) of cost-causing activities that can be performed for operational purposes in cost centers. The activities are divided up (classified) into activity types in a cost-oriented manner, that is, with consideration given to the cost structure existing for the activities. Activity types are differentiated according to their allocatability, the manner in which activity quantities are recorded, and also according to whether fixed costs are allocated in internal activity-based allocations. To a cost center several activity types can be assigned for which activities can be performed in the cost center. Activity types can be divided up (classified), for example, for reporting purposes. For the activity type groupings, a multi-level (if required) hierarchy can be created that specifies how the groupings are grouped together or refined level-by-level.

Application Area: Controlling

BAPI: ActivityType.GetList

Documentation: You can use this method, using selection criteria for a given day, to provide a list of all activity types (controlling area, activity type, and activity type name) matching the criteria. You can enter the criteria as a single controlling area value, as controlling area intervals, or as an upper limit controlling area. If you do not enter any criteria, you will retrieve a list of all activity types. You may also enter a search string with a masking symbol (* or +). The method selects all activity types using the string in their names, titles, or descriptions. It returns a list of all activity types if you do not enter any restrictions. If the R/3 System does not find any data, it raises an exception.

Availability: R/3 3.1G

BAPI: ActivityType.GetPrices

Documentation: This method determines activity type prices for cost center/activity type combinations for a given day. The activity types receiving the prices are given in the form of a table with object Ids (controlling area, activity type). You can enter the cost centers by giving a single cost center value only, an interval of cost centers, or an upper limit cost center number only. If you do not enter these selections, the method searches for all activity types in the object table for cost centers. You determine the controlling area by entries in the object table. The controlling areas of different object entries may vary, but the fiscal year variants of the controlling areas must be compatible so that the R/3 System can deal with the valuation using the same period and fiscal year structures. If this is not the case, the R/3 System raises an exception. The method returns all cost center/activity type combinations fitting the given selection criteria, in addition to the corresponding activity type prices for the given date' s period. In addition to cost center/activity type texts, the total activity price per quantity unit, the activity unit, and the activity price unit in controlling area currency are returned. If no data can be found, the R/3 System raises an exception.

Availability: R/3 3.1G

Business Object: APAccount

Documentation: Documentation currently not available

Application Area: Financial Accounting

BAPI: APAccount.GetBalancedItems

Documentation: This method lists the clearing transactions carried out for a vendor account in a given period. Clearing entries (for example payments) and the items to which they relate (for example invoices, credit memos) are displayed together.

Availability: R/3 3.1G

BAPI: APAccount.GetCurrentBalance

Documentation: This function module displays a vendor' s balance for the current fiscal year. The following information is displayed: 1. the balance for standard transactions and 2. the total balance of standard transactions and all general ledger transactions.

Availability: R/3 3.1G

BAPI: APAccount.GetKeyDateBalance

Documentation: This function module provides a vendor' s total balance on a given key date. This total balance contains both standard and special general ledger transactions. If you so wish, noted items can also be included in this balance. The balance is displayed per transaction currency. If you so wish, the balance can be broken down further per special general ledger transaction.

Availability: R/3 3.1G

BAPI: APAccount.GetOpenItems

Documentation: This function module lists a vendor' s open items at a given key date. If you so wish, noted items can also be selected for display.

Availability: R/3 3.1G

BAPI: APAccount.GetPeriodBalances

Documentation: This function module displays a vendor' s current balance for the current fiscal year, together with the transactions and purchases per period. It also displays the annual values of special general ledger transactions in a table.

Availability: R/3 3.1G

BAPI: APAccount.GetStatement

Documentation: This function module lists the postings made to a vendor account in a given period. If you so wish, noted items can also be included.

Availability: R/3 3.1G

Business Object: Applicant

Documentation: The business object Applicant is a person within or outside the organization who uses an application to express interest in entering into a work relationship or changing an existing work relationship.

Application Area: Personnel Management

BAPI: Applicant.ChangePassword

Documentation: You can use this method to change the password assigned to an applicant. The password is stored in encrypted form.

Availability: R/3 3.1G

BAPI: Applicant.CheckPassword

Documentation: You can use this method to check the password assigned to an applicant.

Availability: R/3 3.1G

BAPI: Applicant.CreateFromData

Documentation: You can use this method to create an applicant. The system carries out all the consistency checks performed when entering applicant data using the R/3 transactions. An applicant is created only if all checks have been performed successfully. The system returns the applicant number it has assigned and a return code.

Availability: R/3 3.1G

BAPI: Applicant.CreatePassword

Documentation: You can use this method to create a password entry for an applicant. The password itself has to be created with method Applicant.InitPassword.

Availability: R/3 3.1G

BAPI: Applicant.DeletePassword

Documentation: You can use this method to delete an entry for an applicant.

Availability: R/3 3.1G

BAPI: Applicant.ExistenceCheck

Documentation: You can use this method to check whether an applicant exists.

Availability: R/3 3.1G

BAPI: Applicant.GetPassword

Documentation: This method enables status information concerning password management for an applicant to be read.

Availability: R/3 3.1G

BAPI: Applicant.GetStatus

Documentation: You can use this method to determine an applicant' s overall status and the status of his/her vacancy assignments.

Availability: R/3 3.1G

BAPI: Applicant.InitPassword

Documentation: You can use this method to initialize the password assigned to an applicant. An initial password is returned. The password itself is stored in encrypted form.

Availability: R/3 3.1G

Business Object: ARAccount

Documentation: Documentation currently not available

Application Area: Financial Accounting

BAPI: ARAccount.GetBalancedItems

Documentation: This function module lists the clearing transactions made to a customer account in a given period. Clearing entries (for example, payments) and the items that they cleared (for example invoices, credit memos) are displayed together.

Availability: R/3 3.1G

BAPI: ARAccount.GetCurrentBalance

Documentation: This function module provides a customer' s balance for the current fiscal year. It displays both the balance of standard transactions and the total balance of standard transactions and all special general ledger transactions.

Availability: R/3 3.1G

BAPI: ARAccount.GetKeyDateBalance

Documentation: This function module supplies a customer' s total balance at a given key date. The total balance includes both standard and special general ledger transactions. If you so wish, noted items can also be included in this balance. The balance is displayed per transaction currency but can can be further broken down per special general ledger transaction.

Availability: R/3 3.1G

BAPI: ARAccount.GetOpenItems

Documentation: This function module lists a customer' s open items at a given key date. If you so wish, noted items can also be displayed.

Availability: R/3 3.1G

BAPI: ARAccount.GetPeriodBalances

Documentation: This function module lists a customer' s current balance for the current fiscal year, together with the transactions and sales per period. The annual values for special general ledger transactions are also displayed in a table.

Availability: R/3 3.1G

BAPI: ARAccount.GetStatement

Documentation: This function module lists the postings made to a customer account in a given period. If you so wish, noted items can also be selected for display.

Availability: R/3 3.1G

Business Object: Attendee

Documentation: Documentation currently not available

Application Area: Training and Event Management

BAPI: Attendee.ChangePassword

Documentation: You can use this method to change the password assigned to an attendee. The password itself is stored in encrypted form.

Availability: R/3 3.1G

BAPI: Attendee.CheckExistence

Documentation: You can use this method to establish whether an attendee exists (for example, a personnel number in HR master data or a customer in the customer master). If you specify a start and end date, a check is conducted to see if the attendee exists in this period. If the attendee does not exist in the period specified, this triggers an exception. If the attendee does exist in the period specified, the attendee' s validity period is also returned.

Availability: R/3 3.1G

BAPI: Attendee.CheckPassword

Documentation: You can use this method to verify the password assigned to an attendee. The password is stored in encrypted form.

Availability: R/3 3.1G

BAPI: Attendee.GetBookList

Documentation: You can use this method to retrieve the business events booked for an attendee within a given period.

Availability: R/3 3.1G

BAPI: Attendee.GetCompanyBookList

Documentation: You can use this method to retrieve the business events booked for a group attendee within a specific period. A group attendee is, for example, a customer who can make bookings for either unnamed attendees (N.N. bookings) or named attendees (individual bookings).

Availability: R/3 3.1G

BAPI: Attendee.GetCompanyPrebookList

Documentation: You can use this method to retrieve a group attendee' s prebookings on business event types for a specific selection period. A group attendee is, for example, a customer who can make prebookings for either unnamed attendees (N.N. prebookings) or named attendees (individual prebookings).

Availability: R/3 3.1G

BAPI: Attendee.GetPrebookList

Documentation: You can use this method to retrieve an attendee' s prebookings on business event types within a given period.

Availability: R/3 3.1G

BAPI: Attendee.GetTypeList

Documentation: You can use this method to retrieve attendee types that were flagged as being Internet attendee types in Training and Event Management Customizing.

Availability: R/3 3.1G

Business Object: BapiService

Documentation: Documentation currently not available

Application Area: Cross-Application Components

BAPI: BapiService.DataConversionExt2Int

Documentation: This method converts data from external into internal format

Availability: R/3 4.0A

BAPI: BapiService.DataConversionInt2Ext

Documentation: This method converts data from internal into external format

Availability: R/3 4.0A

Business Object: BusinessArea

Documentation: The business object Business area is the organizational unit in external accounting that corresponds to a selected area of activity or responsibility within an organization to which the value movements entered in financial accounting can be assigned.

Application Area: Financial Accounting

BAPI: BusinessArea.ExistenceCheck

Documentation: This method ascertains if the business area specified exists in the system. The result of the check is recorded in a return code.

Availability: R/3 3.0E

BAPI: BusinessArea.GetDetail

Documentation: This method enables the user to access further information about a business area. A return code and details determined by the system are returned.

Availability: R/3 4.0A

BAPI: BusinessArea.GetList

Documentation: This method provides the user with a list of available business areas and their description.

Availability: R/3 4.0A

Business Object: BusinessEvent

Documentation: Documentation currently not available

Application Area: Training and Event Management

BAPI: BusinessEvent.GetInfo

Documentation: You can use this method to retrieve various types of information about a business event. This information might include:

- a description of the content of the business event
- the business event organizer the business event costs and price
- the resources that have been reserved, and so on.

Availability: R/3 3.1G

BAPI: BusinessEvent.GetLanguage

Documentation: You can use this method to retrieve all available business event languages.

Availability: R/3 3.1G

BAPI: BusinessEvent.GetSchedule

Documentation: You can use this method to retrieve the time schedule (in days and hours) for a given business event.

Availability: R/3 3.1G

BAPI: BusinessEvent.Init

Documentation: You can use this method to retrieve the default values for selecting business event dates. The criteria used to restrict the selection of business event dates are as follows:

- plan version
- business event language
- business event location
- start date of selection period
- end date of selection period

Availability: R/3 3.1H

Business Object: BusinessEventGroup

Documentation: Documentation currently not available

Application Area: Training and Event Management

BAPI: BusinessEventGroup.GetEventtypeList

Documentation: You can use this method to retrieve a list of business event types for a given business event group.

Availability: R/3 3.1G

BAPI: BusinessEventGroup.GetList

Documentation: You can use this method to retrieve the business event group hierarchy. You can specify a business event group to start from, and then all its subordinate business event groups will be retrieved. If you do not specify a business event group, all business event groups and their relationships to each other in the hierarchy will be retrieved.

Availability: R/3 3.1G

Business Object: BusinessEventtype

Documentation: Documentation currently not available

Application Area: Training and Event Management

BAPI: BusinessEventtype.GetEventList

Documentation: You can use this method to retrieve a list of business event dates for a given business event type within a given selection period.

Availability: R/3 3.1G

BAPI: BusinessEventtype.GetInfo

Documentation: You can use this method to retrieve various types of information about a business event type. This information might include:

- a description of the content of the business event
- the business event organizer
- the business event costs and price
- the qualifications imparted upon completion of the business event,
- the qualifications required to attend the business event, and so on.

Availability: R/3 3.1G

Business Object: BusPartnerEmployee

Documentation: The object business partner employee is a member in the organization of a business partner (vendor, customer, sales partner or competitor).

Application Area: Cross-Application Components

BAPI: BusPartnerEmployee.ChangePassword

Documentation: By using this method you can change the password of a business partner employee. The password is stored in an encrypted form.

Availability: R/3 3.1G

BAPI: BusPartnerEmployee.CheckExistence

Documentation: Using this method you can determine the existence of a business partner employee.

Availability: R/3 3.1G

BAPI: BusPartnerEmployee.CheckPassword

Documentation: Using this method, you can check the password of a business partner employee. The password is stored in an encrypted form. The methods returns notifications.

Availability: R/3 3.1G

BAPI: BusPartnerEmployee.CreatePassword

Documentation: Using this method you can create a business partner employee entry in the password management. After you have successfully carried out this method, you can assign a password to the business partner employee by using the method BusPartnerEmployee.InitPassword.

Availability: R/3 3.1G

BAPI: BusPartnerEmployee.DeletePassword

Documentation: Using this method you can delete the password for a business partner employee.

Availability: R/3 3.1G

BAPI: BusPartnerEmployee.GetPassword

Documentation: Using this method, you can read status information on a business partner employee.

Availability: R/3 3.1G

BAPI: BusPartnerEmployee.InitPassword

Documentation: Using this method, you can assign an initial password to a business partner employee. The password is stored in an encrypted form. Note that you must successfully invoke the method BusPartnerEmployee.CreatePassword in advance.

Availability: R/3 3.1G

Business Object: Company

Documentation: The business object Company is the smallest organizational unit for which an individual financial statement is drawn up according to a given country' s commercial code.

Application Area: Financial Accounting

BAPI: Company.ExistenceCheck

Documentation: This method ascertains if the company specified exists in the system. The result of the check is recorded in a return code.

Availability: R/3 3.1G

BAPI: Company.GetDetail

Documentation: This method enables the user to access further information about a company. The company' s detailed data as determined by the system is returned. Any problems arising are returned as a return code message.

Availability: R/3 3.1G

BAPI: Company.GetList

Documentation: This method provides the user with a list of companies. The system returns both the keys and the names of the companies, in so far as they exist in the system. Problems arising are returned as a return code message.

Availability: R/3 3.1G

Business Object: CompanyCode

Documentation: The business object Company code is the smallest organizational unit of financial accounting for which a complete, self-contained accounting can be represented. This contains the recording of all accountable events and the drawing up of all statements for individual accounts as required by law, such as balance sheets as well as profit and loss statements.

Application Area: Financial Accounting

BAPI: CompanyCode.ExistenceCheck

Documentation: This method ascertains if the company code specified exists in the system. The result of the check is recorded in a return code.

Availability: R/3 4.0A

BAPI: CompanyCode.GetDetail

Documentation: This method enables the user to access further information for a company code. The system returns the detail data and address data for the company code. Problems arising are returned as a return code message.

Availability: R/3 3.1G

BAPI: CompanyCode.GetList

Documentation: This method provides the user with a list of company codes. The system returns both the keys and the names of the company codes, in so far as the latter exist in the system. Problems arising are returned as a return code message.

Availability: R/3 3.1G

BAPI: CompanyCode.GetPeriod

Documentation: This method returns the associated fiscal year for a company code and posting date.

Availability: R/3 4.0A

Business Object: ControllingArea

Documentation: The business object Controlling area is the organizational unit within the enterprise for which a complete, self-contained cost accounting can be carried out. Company codes and controlling areas are assigned to one another. For the implementation of cross-company code cost accounting, a controlling area can group together several company codes. It can also be identical with a company code. For the purposes of a common profitability analysis several controlling areas can be assigned to an operating result area.

Application Area: Controlling

BAPI: ControllingArea.GetDetail

Documentation: Supplies the following details about a controlling area:

- Name
- Currency
- Chart of accounts
- Fiscal year variant

Availability: R/3 3.1G

BAPI: ControllingArea.Getlist

Documentation: This method offers a list of existing controlling areas. You can find additional information for an individual controlling area by invoking the method ControllingArea.GetDetail

Availability: R/3 3.1G

BAPI: ControllingArea.GetPeriod

Documentation: Documentation currently not available

Availability: R/3 4.0A

BAPI: ControllingArea.GetPeriodLimits

Documentation: Documentation currently not available

Availability: R/3 4.0A

BAPI: ControllingArea.Find

Documentation: Documentation currently not available

Availability: R/3 4.0A

Business Object: ControllingDocument

Documentation: Documentation currently not available

Application Area: Controlling

BAPI: ControllingDocument.GetDetail

Documentation: Documentation currently not available

Availability: R/3 3.1G

Business Object: CostCenter

Documentation: The business object Cost center contains the organizational unit within a controlling area that represents a clearly delimited place from which costs originate. The delimitation can be carried out according to functional, accounting, spatial and/or cost responsibility viewpoints. A cost center is assigned to exactly one company code and, if required, to a business area.

Application Area: Controlling

BAPI: CostCenter.GetDetail

Documentation: This method supplies details of a cost center. The method requires the controlling area, the cost center, and a date. The method returns the cost center manager and the structures with the address and communication data. If the cost center is not found, the R/3 System raises an exception.

Availability: R/3 3.1G

BAPI: CostCenter.GetList

Documentation: You can use this method, to provide a list of all cost centers (controlling area, cost center, and cost center name) that match the selection criteria for a given day. You can enter the criteria as a single company code, an upper limit company code or as a company code interval. If you do not enter any criteria, you will retrieve a list of all cost centers. If the R/3 System does not find any data, it raises an exception.

Availability: R/3 3.1G

Business Object: CostEstimate

Documentation: The business object Cost estimate contains costs for a costing object that accrue in connection with the production and the sales of goods or services. A cost estimate is a detailed list of the costs of a costing object. A cost estimate is determined by the costing type and the costing valuation variant. A cost estimate can be related or unrelated, that is, based on existing quantity and value structures such as bills of material and routings, or on cost specifications defined by the user. A cost estimate consists of several items that contain the determined cost components (for example, material costs, personnel costs and so on). Furthermore, to a cost estimate belong several itemizations in which the costs are contained for each resource used.

Application Area: Controlling

BAPI: CostEstimate.GetDetail

Documentation: You can use this method to access cost estimates for materials in the R/3 System. You receive the following data from the costing header:

- Currency
- Material

- Plant
- Valid from
- Valid to
- Valuation date
- Quantity structure date
- Costing status
- Costing lot size
- Costing variant
- Material with BOM (indicator 01)
- Material without BOM (indicator 02)

You receive the following data based on the cost component split:

- Fixed costs
- Total costs

You receive the following data based on the cost component split and a certain cost component:

- Fixed costs
- Total costs

Availability: R/3 4.0A

BAPI: CostEstimate.GetExplosion

Documentation: This method can be used to explode the bill of material (BOM) structure used as a basis for the cost estimate. It allows the user to expand the structure level by level. The costed multilevel BOM is a hierarchical overview of all items of a costed material. It uses the itemization to break down the total costs of the product, thus allowing you to see the input quantity and costs for every material in the BOM at every level. Any problems arising are returned as a return code message.

Availability: R/3 4.0A

BAPI: CostEstimate.GetItemization

Documentation: This method enables the user to display the itemization for the cost estimates selected.

The itemization provides detailed information on each costing item about the origin of the costs for each transaction in the routing of the material costed. Any problems arising are returned as a return code message.

Availability: R/3 4.0A

BAPI: CostEstimate.GetList

Documentation: This method enables you to access cost estimates for materials in the R/3 System. The selection provides you with the following data:

- Reference object
- Cost estimate number
- Costing type
- Costing date
- Costing version
- Valuation variant
- Manual cost estimate
- Cost estimate valid to
- Costing status
- Material with BOM (indicator 01)
- Material without BOM (indicator 02)

Availability: R/3 4.0A

Business Object: CostType

Documentation: Documentation currently not available

Application Area: Controlling

BAPI: CostType.GetFixaccount

Documentation: Documentation currently not available

Availability: R/3 3.1G

BAPI: CostType.GetFixaccountlist

Documentation: Documentation currently not available

Availability: R/3 3.1G

Business Object: Creditor

Documentation: The business object Creditor is a business partner to whom payables are owed for goods delivered or services rendered.

Application Area: Financial Accounting

BAPI: Creditor.ChangePassword

Documentation: This method enables the user to change a creditor password.

Availability: R/3 3.1G

BAPI: Creditor.CheckPassword

Documentation: This method enables a password for a creditor to be checked.

Availability: R/3 3.1G

BAPI: Creditor.CreatePasswordRegistry

Documentation: This function module is used to create an entry for a vendor in password management. A password is not generated for the vendor.

Availability: R/3 3.1G

BAPI: Creditor.DeletePasswordRegistry

Documentation: This function module enables an entry for a vendor to be deleted from password management.

Availability: R/3 3.1G

BAPI: Creditor.ExistenceCheck

Documentation: This method enables the user to establish whether a creditor is defined. Where a company code is stated, an additional check is made as to whether the creditor is created in this company code.

Availability: R/3 3.1G

BAPI: Creditor.Find

Documentation: This method selects all creditors according to predefined selection criteria.

Availability: R/3 4.0A

BAPI: Creditor.GetDetail

Documentation: This method enables the retrieval of detailed creditor data. The following data is provided: general data, a table of bank data and, where a company code is given, company code specific data.

Availability: R/3 3.1G

BAPI: Creditor.GetPasswordRegistry

Documentation: This function module enables status information concerning password management for a vendor to be read.

Availability: R/3 3.1H

BAPI: Creditor.InitPassword

Documentation: This method enables a creditor password to be initialized. An initial password is returned. Use method Creditor. CreatePassword in advance.

Availability: R/3 3.1G

Business Object: Currency

Documentation: Documentation currently not available

Application Area: Financial Accounting

BAPI: Currency.GetDecimals

Documentation: Documentation currently not available

Availability: R/3 4.0A

Business Object: Customer

Documentation: The business object Customer is a business partner with whom business relations can exist with regard to the issuing of materials and/or the rendering of services.

Application Area: Financial Accounting

BAPI: Customer.ChangeFromData

Documentation: This method is used to change the address data of a customer in the R/3 system.

Availability: R/3 4.0A

BAPI: Customer.ChangePassword

Documentation: This method allows you to change the password of a customer.

Availability: R/3 3.1H

BAPI: Customer.CheckExistence

Documentation: This method checks whether a specific customer is known.

Availability: R/3 3.1H

BAPI: Customer.CheckPassword

Documentation: Using this method you can check a customer' s password. The password is stored in an encrypted form.

Availability: R/3 3.1H

BAPI: Customer.CreateFromData

Documentation: This method is used to create a customer in the R/3 system from address data entered. Company code and sales area-relevant data is updated by a reference customer and sales area data (sales organization, distribution channel and division).

Availability: R/3 4.0A

BAPI: Customer.CreatePassword

Documentation: This method is used to create an entry for a customer in password management. The password itself is not generated. To do so use method Customer.InitPassword.

Availability: R/3 3.1H

BAPI: Customer.DeletePassword

Documentation: This method deletes a customer from password management.

Availability: R/3 3.1H

BAPI: Customer.GetPassword

Documentation: This method returns status information concerning password management for a specific customer.

Availability: R/3 3.1H

BAPI: Customer.GetSalesAreas

Documentation: This method reads all the sales areas for a customer.

Availability: R/3 3.1G

BAPI: Customer.InitPassword

Documentation: This method enables a customer password to be initialized. An initial password is returned. Use method Customer.CreatePassword in advance.

Availability: R/3 3.1H

BAPI: Customer.Search

Documentation: Documentation currently not available

Availability: R/3 4.0A

Business Object: CustomerInquiry

Documentation: Documentation currently not available

Application Area: Sales and Distribution

BAPI: CustomerInquiry.CreateFromData

Documentation: You can use this method to create customer inquiries.

Availability: R/3 4.0A

Business Object: CustomerQuotation

Documentation: Documentation currently not available

Application Area: Sales and Distribution

BAPI: CustomerQuotation.CreateFromData

Documentation: You can create customer quotations with this method.

Availability: R/3 4.0A

Business Object: Debtor

Documentation: The business object Debtor is a business partner from whom a receivable is due for goods delivered or services rendered.

Application Area: Financial Accounting

BAPI: Debtor.ChangePassword

Documentation: This method enables the user to change a debtor password.

Availability: R/3 3.1G

BAPI: Debtor.CheckPassword

Documentation: This method enables a debtor password to be checked.

Availability: R/3 3.1G

BAPI: Debtor.CreatePasswordRegistry

Documentation: This function module enables you to create an entry for a customer in password management. It does not generate a password for a customer.

Availability: R/3 3.1G

BAPI: Debtor.DeletePasswordRegistry

Documentation: This function module is used to delete an entry for a customer in password management.

Availability: R/3 3.1G

BAPI: Debtor.ExistenceCheck

Documentation: This method enables you to check whether a debtor exists. Where a company code is given an additional check is made as to whether the debtor is created in this company code.

Availability: R/3 3.1G

BAPI: Debtor.Find

Documentation: This method selects all debtors according to predefined selection criteria.

Availability: R/3 4.0A

BAPI: Debtor.GetDetail

Documentation: This method enables the retrieval of detailed debtor data. It provides general data, a table of bank data and, where a company code has been given, data specific to a company code.

Availability: R/3 3.1G

BAPI: Debtor.GetPasswordRegistry

Documentation: This function module enables you to read status information relating to password management for a customer.

Availability: R/3 3.1H

BAPI: Debtor.InitPassword

Documentation: This method enables you to initialize a debtor password. An initial password is returned.

Availability: R/3 3.1G

Business Object: DebtorCreditAccount

Documentation: Documentation currently not available

Application Area: Financial Accounting

BAPI: DebtorCreditAccount.GetDetail

Documentation: Documentation currently not available

Availability: R/3 4.0A

BAPI: DebtorCreditAccount.GetHighestDunningLevel

Documentation: Documentation currently not available

Availability: R/3 4.0A

BAPI: DebtorCreditAccount.GetOldestOpenItem

Documentation: Documentation currently not available

Availability: R/3 4.0A

BAPI: DebtorCreditAccount.GetOpenItemsStructure

Documentation: Documentation currently not available

Availability: R/3 4.0A

BAPI: DebtorCreditAccount.GetStatus

Documentation: Documentation currently not available

Availability: R/3 4.0A

BAPI: DebtorCreditAccount.ReplicateStatus

Documentation: Documentation currently not available

Availability: R/3 4.0A

Business Object: DistributionModel

Documentation: Documentation currently not available

Application Area: Cross-Application Components

BAPI: DistributionModel.GetInfo

Documentation: This method enables you to read all the data in the distribution model.

Availability: R/3 4.0A

Business Object: EmpBenefitHealthPlan

Documentation: Documentation currently not available

Application Area: Personnel Management

BAPI: EmpBenefitHealthPlan.GetDependents

Documentation: This function module is used as the basis for the BAPI method EmpBenefitHealthPlan.GetDependents. This method returns a list of the dependents an employee has included in benefit plan enrollments. It reads the employee' s health plan enrollments on a specified date, and returns the list of dependents.

Availability: R/3 4.0A

BAPI: EmpBenefitHealthPlan.GetPossDependents

Documentation: This function module is used as the basis for the BAPI method EmpBenefitHealthPlan.GetPossDependents. This method returns a list of the dependents an employee may include in health plan enrollments on a specified date.

Availability: R/3 4.0A

Business Object: EmpBenefitInsurePlan

Documentation: Documentation currently not available

Application Area: Personnel Management

BAPI: EmpBenefitInsurePlan.GetBeneficiaries

Documentation: This function module is used as the basis for the BAPI method EmpBenefitInsurePlan.GetBeneficiaries. This method returns a list of the beneficiaries an employee has included in benefit plan enrollments. It reads the employee' s insurance plan enrollments on a specified date, and returns the list of beneficiaries.

Availability: R/3 4.0A

BAPI: EmpBenefitInsurePlan.GetPossBeneficiaries

Documentation: This function module is used as the basis for the BAPI method EmpBenefitInsurePlan.GetPossBeneficiaries. This method returns a list of the beneficiaries an employee may include in insurance plan enrollments on a specified date.

Availability: R/3 4.0A

Business Object: EmpBenefitMiscelPlan

Documentation: Documentation currently not available

Application Area: Personnel Management

BAPI: EmpBenefitMiscelPlan.GetBeneficiaries

Documentation: This function module is used as the basis for the BAPI method EmpBenefitMiscelPlan.GetBeneficiaries. This method returns a list of the beneficiaries an employee has included in benefit plan enrollments. It reads the employee' s miscellaneous plan enrollments on a specified date, and returns the list of beneficiaries.

Availability: R/3 4.0A

BAPI: EmpBenefitMiscelPlan.GetDependents

Documentation: This function module is used as the basis for the BAPI method EmpBenefitMiscelPlan.GetDependents. This method returns a list of the dependents an employee has included in benefit plan enrollments. It reads the employee' s miscellaneous plan enrollments on a specified date, and returns the list of dependents.

Availability: R/3 4.0A

BAPI: EmpBenefitMiscelPlan.GetInvestments

Documentation: This function module is used as the basis for the BAPI method EmpBenefitMiscelPlan.GetInvestments. This method returns a list of the investments an employee has selected in benefit plan enrollments. It reads the employee' s miscellaneous plan enrollments on a specified date, and returns the list of investments.

Availability: R/3 4.0A

BAPI: EmpBenefitMiscelPlan.GetPossBeneficiaries

Documentation: This function module is used as the basis for the BAPI method EmpBenefitMiscelPlan.GetPossBeneficiaries. This method returns a list of the beneficiaries an employee may include in miscellaneous plan enrollments on a specified date.

Availability: R/3 4.0A

BAPI: EmpBenefitMiscelPlan.GetPossDependents

Documentation: This function module is used as the basis for the BAPI method EmpBenefitMiscelPlan.GetPossDependents. This method returns a list of the dependents an employee may include in miscellaneous plan enrollments on a specified date.

Availability: R/3 4.0A

BAPI: EmpBenefitMiscelPlan.GetPossInvestments

Documentation: This function module is used as the basis for the BAPI method EmpBenefitMiscelPlan.GetPossInvestments. This method returns a list of the investments an employee may select in miscellaneous plan enrollments on a specified date.

Availability: R/3 4.0A

Business Object: EmpBenefitSavingPlan

Documentation: Documentation currently not available

Application Area: Personnel Management

BAPI: EmpBenefitSavingPlan.GetBeneficiaries

Documentation: This function module is used as the basis for the BAPI method EmpBenefitSavingPlan.GetBeneficiaries. This method returns a list of the beneficiaries an employee has included in benefit plan enrollments. It reads the employee' s savings plan enrollments on a specified date, and returns the list of beneficiaries.

Availability: R/3 4.0A

BAPI: EmpBenefitSavingPlan.GetInvestments

Documentation: This function module is used as the basis for the BAPI method EmpBenefitSavingPlan.GetInvestments. This method returns a list of the investments an employee has selected in benefit plan enrollments. It reads the employee' s savings plan enrollments on a specified date, and returns the list of investments.

Availability: R/3 4.0A

BAPI: EmpBenefitSavingPlan.GetPossBeneficiaries

Documentation: This function module is used as the basis for the BAPI method EmpBenefitSavingPlan.GetBeneficiaries. This method returns a list of the beneficiaries an employee may include in savings plan enrollments on a specified date.

Availability: R/3 4.0A

BAPI: EmpBenefitSavingPlan.GetPossInvestments

Documentation: This function module is used as the basis for the BAPI method EmpBenefitSavingPlan.GetPossInvestments. This method returns a list of the investments an employee may select in savings plan enrollments on a specified date.

Availability: R/3 4.0A

Business Object: Employee

Documentation: The business object Employee is a person who contributes or has contributed to the creation of goods and services within the enterprise. As a rule, this occurs on the basis of a work contract or a contract for services. The employee constitutes the main focus of interest for the majority of human resource subareas (personnel administration, time management, payroll accounting, etc.). All business processes within these areas relate to the employee.

Application Area: Personnel Management

BAPI: Employee.ChangePassword

Documentation: You can use this method to change an employee' s password.

Availability: R/3 4.0A

BAPI: Employee.CheckPassword

Documentation: You can use this method to check an employee' s password.

Availability: R/3 4.0A

BAPI: Employee.CreatePassword

Documentation: By using this method, an entry for an employee is created. A password is not generated for the employee. To do so use method Employee.InitPassword.

Availability: R/3 4.0A

BAPI: Employee.DeletePassword

Documentation: By using this method, you can delete the password of an employee.

Availability: R/3 4.0A

BAPI: Employee.Dequeue

Documentation: This method unlocks all data of one employee.

Availability: R/3 4.0A

BAPI: Employee.Enqueue

Documentation: This method locks all data of one employee.

Availability: R/3 4.0A

BAPI: Employee.GetList

Documentation: This method provides data on the employees that fulfill the specified search criteria (information such as: Organizational assignment, personal data, internal control, communication, document from optical archive). The system only retrieves those employees that fulfil the selection criteria. If you do not enter any restriction conditions the selection will not take place.

Availability:R/3 4.0A

BAPI: Employee.GetPassword

Documentation: By using this method, you can read the status information from the password management for a specific employee.

Availability: R/3 4.0A

BAPI: Employee.InitPassword

Documentation: By using this method you can initialize a password for an employee. An initial password is given. The coded password is stored.

Availability: R/3 4.0A

Business Object: EmployeeAbsence

Documentation: The business object EmployeeAbsence stores information on an employee´s absences.

Application Area: Personnel Time Management

BAPI: EmployeeAbsence.Approve

Documentation: With this method a previous request for an absence (EmployeeAbsence.Request) can be approved.

Availability: R/3 4.0A

BAPI: EmployeeAbsence.Change

Documentation: This method can be used to change an absence entry.

Availability: R/3 4.0A

BAPI: EmployeeAbsence.Create

Documentation: This method creates a new absence entry for an employee.

Availability: R/3 4.0A

BAPI: EmployeeAbsence.Delete

Documentation: This method deletes an absence entry.

Availability: R/3 4.0A

BAPI: EmployeeAbsence.GetDetail

Documentation: This method shows all information stored for one absence.

Availability: R/3 4.0A

BAPI: EmployeeAbsence.GetList

Documentation: This method retrieves the whole absence entry list of an employee.

Availability: R/3 4.0A

BAPI: EmployeeAbsence.Request

Documentation: This methods requests an absence for an employee, which can be approved later using EmployeeAbsence.Approve.

Availability: R/3 4.0A

BAPI: EmployeeAbsence.SimulateCreation

Documentation: Using this method, the process of creating an Absences infotype record can be simulated. All required consistency checks are carried out; the record is not written to the database, however.

Availability: R/3 4.0A

Business Object: EmployeeAbstract

Documentation: Documentation currently not available

Application Area: Personnel Management

BAPI: EmployeeAbstract.ChangePassword

Documentation: You can use this method to change an employee' s password.

Availability: R/3 3.1G

BAPI: EmployeeAbstract.CheckPassword

Documentation: You can use this method to check an employee' s password.

Availability: R/3 3.1G

BAPI: EmployeeAbstract.CreatePassword

Documentation: By using this method, an entry for an employee is created. A password is not generated for the employee.

Availability: R/3 3.1G

BAPI: EmployeeAbstract.DeletePassword

Documentation: By using this method, you can delete the entry for an employee.

Availability: R/3 3.1G

BAPI: EmployeeAbstract.Dequeue

Documentation: You can use the DEQUEUE method to unlock an employee so that the records stored for this person can be accessed. If an employee is locked using the ENQUEUE method, the user who set the lock can access this employee' s records. Other users are denied access to these records. The DEQUEUE method removes the lock.

Availability: R/3 3.1G

BAPI: EmployeeAbstract.Enqueue

Documentation: You can use this method to lock an employee so that the records stored for this person cannot be accessed. When an employee is locked, only the user who has set the lock can access the records for this employee. Other users are denied access. Only when an employee is locked for other users, can his/her records be maintained (created, changed, deleted, copied) by the person who has set the lock.

Availability: R/3 3.1G

Business Object: EmployeeBankDetail

Documentation: This object holds an employee´s bank details history.

Application Area: Personnel Management

BAPI: EmployeeBankDetail.Approve

Documentation: With this method a previous request for a change of bank details (EmployeeBankDetail.Request) can be approved.

Availability: R/3 4.0A

BAPI: EmployeeBankDetail.Change

Documentation: This method can be used to change a bank details entry.

Availability: R/3 4.0A

BAPI: EmployeeBankDetail.Create

Documentation: This method creates a new bank details entry for an employee.

Availability: R/3 4.0A

BAPI: EmployeeBankDetail.CreateSuccessor

Documentation: This method creates a new bank details entry for an employee and copies all non-specified data fields from the previous entry.

Availability: R/3 4.0A

BAPI: EmployeeBankDetail.Delete

Documentation: This method deletes a bank details entry.

Availability: R/3 4.0A

BAPI: EmployeeBankDetail.GetDetail

Documentation: This method shows all information stored for one bank details entry.

Availability: R/3 4.0A

BAPI: EmployeeBankDetail.GetList

Documentation: This method retrieves the all bank details entries of an employee.

Availability: R/3 4.0A

BAPI: EmployeeBankDetail.Request

Documentation: This methods requests a change of bank details for an employee, which can be approved later using EmployeeBankDetail.Approve.

Availability: R/3 4.0A

BAPI: EmployeeBankDetail.Simulatecreation

Documentation: You can use this method to simulate creation of a Bank Details record. During the simulation all necessary consistency checks are performed, but the record is not written to the database.

Availability: R/3 4.0A

Business Object: EmployeeBasicPay

Documentation: The business object EmployeeBasicPay stores all information about regularly reoccuring payments to an employee. Changes in income, benefits, etc. are stored as successive entries in a list.

Application Area: Personnel Management

BAPI: EmployeeBasicPay.Approve

Documentation: After a new, but non-active entry has been created (using EmployeeBasicPay.Request), this method is used to approve the new entry and to make it active. The end date of the previous entry gets set.

Availability: R/3 4.0A

BAPI: EmployeeBasicPay.Change

Documentation: This method can be used to change basicpay entries.

Availability: R/3 4.0A

BAPI: EmployeeBasicPay.Create

Documentation: This method creates a new payment entry for an employee.

Availability: R/3 4.0A

BAPI: EmployeeBasicPay.CreateSuccessor

Documentation: This method creates a new payment entry for an employee and copies all non-specified data fields from the previous entry. The end date of the previous entry gets set.

Availability: R/3 4.0A

BAPI: EmployeeBasicPay.Delete

Documentation: This method creates a new payment entry for an employee and copies all non-specified data fields from the previous entry. The end date of the previous entry gets set.

Availability: R/3 4.0A

BAPI: EmployeeBasicPay.GetDetail

Documentation: This method displays all information hold by this business object.

Availability: R/3 4.0A

BAPI: EmployeeBasicPay.GetList

Documentation: This method retrieves the whole entry list of an employee, representing his or her payment and benefit history.

Availability: R/3 4.0A

BAPI: EmployeeBasicPay.Request

Documentation: This method creates a new payment entry for an employee and copies all non-specified data fields from the previous entry. However, the new entry is not yet active, but needs to be approved first (using EmployeeBasicPay.Approve).

Availability: R/3 4.0A

BAPI: EmployeeBasicPay.SimulateCreation

Documentation: You can use this method to simulate creation of a Basic Pay record. During the simulation all necessary consistency checks are performed, but the record is not written to the database.

Availability: R/3 4.0A

Business Object: EmployeeBenefit

Documentation: The business object EmployeeBenefit encompasses the set of employer benefits in which an employee is currently participating, or for which an employee is eligible. This objects is especially useful with Employee Self Service providing access to the SAP system using alternative presentation mediums such as Web, Kiosks, Interactive Voice Response, and the like.

Application Area: Personnel Management

BAPI: EmployeeBenefit.GetEventList

Documentation: This method provides a list of benefits events which are currently active for an employee. Active benefits events allow an employee to perform changes to their benefits participation elections which are not generally allowed.

Availability: R/3 4.0A

BAPI: EmployeeBenefit.GetOpenEnrollmentPeriod

Documentation: This method returns the dates of the period in which an employee is able to make any and all changes to their benefits participation elections.

Availability: R/3 4.0A

BAPI: EmployeeBenefit.GetOffer

Documentation: This method provides a benefits offer which consists of those benefits and changes which an employee is allowed to make at the specific point and time. This offer considers whether there there are active benefits events for an employee and whether or not it is an open enrollment period, as well as the participation eligibility rules associated with the particular benefits.

Availability: R/3 4.0A

BAPI: EmployeeBenefit.GetParticipation

Documentation: This method returns a set of benefit plans for which an employee is enrolled at a specific point in time.

Availability: R/3 4.0A

BAPI: EmployeeBenefit.CheckSelection

Documentation: This method performs a check of the benefit elections made by an employee insuring that the employee' s elections is consistent with company rules and policies.

Availability: R/3 4.0A

BAPI: EmployeeBenefit.CreatePlans

Documentation: This method records an employee' s participation in benefit plans.

Availability: R/3 4.0A

BAPI: EmployeeBenefit.DeletePlans

Documentation: This function module is used as the basis for the BAPI method EmployeeBenefit.DeletePlans. This method delimits the employee enrollments on the database. This deletion is done precisely for the period specified.

Availability: R/3 4.0A

BAPI: EmployeeBenefit.GetCorequisitePlans

Documentation: This function module is used as the basis for the BAPI method EmployeeBenefit.GetCorequisitePlans. Certain plans within the complete benefits offer specify other plans in which the employee must also concurrently enroll. These are known as corequisite plans and are delivered by this method. The method uses the plans detailed in the complete benefits offer to fetch corequisites for these plans.

Availability: R/3 4.0A

BAPI: EmployeeBenefit.GetDependants

Documentation: This method returns the set of family and related persons who are currently covered under an employee's benefit plan.

Availability: R/3 4.0A

BAPI: EmployeeBenefit.GetPossDependants

Documentation: This method returns the set of family and related person' which are eligible to be included as dependants for a particular benefit plan.

Availability: R/3 4.0A

Business Object: EmployeeCH

Documentation: Documentation currently not available

Application Area: Personnel Management

BAPI: EmployeeCH.ChangePassword

Documentation: You can use this method to change an employee' s password.

Availability: R/3 3.0E

BAPI: EmployeeCH.CheckPassword

Documentation: You can use this method to check an employee' s password.

Availability: R/3 4.0A

BAPI: EmployeeCH.CreatePassword

Documentation: By using this method, an entry for an employee is created. A password is not generated for the employee. To do so use method Employee.InitPassword.

Availability: R/3 3.0E

BAPI: EmployeeCH.DeletePassword

Documentation: By using this method, you can delete the password of an employee.

Availability: R/3 3.0E

BAPI: EmployeeCH.Dequeue

Documentation: This method unlocks all data of one employee.

Availability: R/3 3.0E

BAPI: EmployeeCH.Enqueue

Documentation: This method locks all data of one employee.

Availability: R/3 3.0E

BAPI: EmployeeCH.GetList

Documentation: This method provides data on the employees that fulfill the specified search criteria (information such as: Organizational assignment, personal data, internal control, communication, document from optical archive). The system only retrieves those employees that fulfil the selection criteria. If you do not enter any restriction conditions the selection will not take place.

Availability: R/3 3.0E

BAPI: EmployeeCH.GetPassword

Documentation: By using this method, you can read the status information from the password management for a specific employee.

Availability: R/3 3.0E

BAPI: EmployeeCH.InitPassword

Documentation: By using this method you can initialize a password for an employee. An initial password is given. The coded password is stored.

Availability: R/3 3.0E

Business Object: EmployeeE

Documentation: Documentation currently not available

Application Area: Personnel Management

BAPI: EmployeeE.ChangePassword

Documentation: You can use this method to change an employee' s password.

Availability: R/3 3.0E

BAPI: EmployeeE.CheckPassword

Documentation: You can use this method to check an employee' s password.

Availability: R/3 3.0E

BAPI: EmployeeE.CreatePassword

Documentation: By using this method, an entry for an employee is created. A password is not generated for the employee. To do so use method Employee.InitPassword.

Availability: R/3 3.0E

BAPI: EmployeeE.DeletePassword

Documentation: By using this method, you can delete the password of an employee.

Availability: R/3 3.0E

BAPI: EmployeeE.Dequeue

Documentation: This method unlocks all data of one employee.

Availability: R/3 3.0E

BAPI: EmployeeE.Enqueue

Documentation: This method locks all data of one employee.

Availability: R/3 3.0E

BAPI: EmployeeE.GetList

Documentation: This method provides data on the employees that fulfill the specified search criteria (information such as: Organizational assignment, personal data, internal control, communication, document from optical archive). The system only retrieves those employees that fulfil the selection criteria. If you do not enter any restriction conditions the selection will not take place.

Availability: R/3 3.0E

BAPI: EmployeeE.GetPassword

Documentation: By using this method, you can read the status information from the password management for a specific employee.

Availability: R/3 3.0E

BAPI: EmployeeE.InitPassword

Documentation: By using this method you can initialize a password for an employee. An initial password is given. The coded password is stored.

Availability: R/3 3.0E

Business Object: EmployeeFamilyMember

Documentation: Documentation currently not available

Application Area: Personnel Management

BAPI: EmployeeFamilyMember.Approve

Documentation: You can use this method to unlock a Family/Related Person record. It enables you to implement the "double verification principle" in conjunction with the REQUEST method. According to this principle, at least two users must take part in writing an active record to the database. One user creates a locked record using the REQUEST method. The other user unlocks or ' activates' this record using the APPROVE method.

Availability: R/3 3.0E

BAPI: EmployeeFamilyMember.Change

Documentation: You can use this method to change a Family/Related Person record.

Availability: R/3 3.0E

BAPI: EmployeeFamilyMember.Create

Documentation: You can use this method to create a Family/Related Person record.

Availability: R/3 3.0E

BAPI: EmployeeFamilyMember.CreateSuccessor

Documentation: You can use this method to create a subsequent Family/Related Person record.

Availability: R/3 3.0E

BAPI: EmployeeFamilyMember.Delete

Documentation: You can use this method to delete a Family/Related Person record.

Availability: R/3 3.0E

BAPI: EmployeeFamilyMember.GetDetail

Documentation: You can use this method to read a Family/Related Person record.

Availability: R/3 3.0E

BAPI: EmployeeFamilyMember.GetList

Documentation: You can use this method to read the keys (instances) of Family/Related Person records which are within the specified time interval.

Availability: R/3 3.0E

BAPI: EmployeeFamilyMember.Request

Documentation: You can use this method to create a locked Family/Related Person record. It enables you to implement the "double verification principle" in conjunction with the APPROVE method. According to this principle, at least two users must take part in writing an active record to the database. One user creates a locked record using the REQUEST method. The other user unlocks or ' activates' this record using the APPROVE method.

Availability: R/3 3.0E

BAPI: EmployeeFamilyMember.Simulatecreation

Documentation: You can use this method to simulate creation of a Family/Related Person record. During the simulation all necessary consistency checks are performed, but the record is not written to the database.

Availability: R/3 3.0E

Business Object: EmployeeGB

Documentation: Documentation currently not available

Application Area: Personnel Management

BAPI: EmployeeGB.ChangePassword

Documentation: You can use this method to change an employee' s password.

Availability: R/3 3.1G

BAPI: EmployeeGB.CheckPassword

Documentation: You can use this method to check an employee' s password.

Availability: R/3 3.1G

BAPI: EmployeeGB.CreatePassword

Documentation: By using this method, an entry for an employee is created. A password is not generated for the employee. To do so use method Employee.InitPassword.

Availability: R/3 3.1G

BAPI: EmployeeGB.DeletePassword

Documentation: By using this method, you can delete the password of an employee.

Availability: R/3 3.1G

BAPI: EmployeeGB.Dequeue

Documentation: This method unlocks all data of one employee.

Availability: R/3 3.1G

BAPI: EmployeeGB.Enqueue

Documentation: This method locks all data of one employee.

Availability: R/3 3.1G

BAPI: EmployeeGB.GetList

Documentation: This method provides data on the employees that fulfill the specified search criteria (information such as: Organizational assignment, personal data, internal control, communication, document from optical archive). The system only retrieves those employees that fulfil the selection criteria. If you do not enter any restriction conditions the selection will not take place.

Availability: R/3 3.1G

BAPI: EmployeeGB.GetPassword

Documentation: By using this method, you can read the status information from the password management for a specific employee.

Availability: R/3 3.1G

BAPI: EmployeeGB.InitPassword

Documentation: By using this method you can initialize a password for an employee. An initial password is given. The coded password is stored.

Availability: R/3 3.1G

Business Object: EmployeeIntControl

Documentation: This object stores room number, phone numbers, car license plates, and other information.

Application Area: Personnel Management

BAPI: EmployeeIntControl.Approve

Documentation: With this method a previous request for an entry can be approved.

Availability: R/3 4.0A

BAPI: EmployeeIntControl.Change

Documentation: This method can be used to change an entry.

Availability: R/3 4.0A

BAPI: EmployeeIntControl.Create

Documentation: This method creates a new entry.

Availability: R/3 4.0A

BAPI: EmployeeIntControl.CreateSuccessor

Documentation: This method creates a new entry and copies all non-specified data fields from the previous entry.

Availability: R/3 4.0A

BAPI: EmployeeIntControl.Delete

Documentation: This method deletes an entry.

Availability: R/3 4.0A

BAPI: EmployeeIntControl.GetDetail

Documentation: This method shows all information stored for one entry.

Availability: R/3 4.0A

BAPI: EmployeeIntControl.GetList

Documentation: This method retrieves the all entries of an employee.

Availability: R/3 4.0A

BAPI: EmployeeIntControl.Request

Documentation: This method creates a new non-active entry, which can be approved later.

Availability: R/3 4.0A

BAPI: EmployeeIntControl.Simulatecreation

Documentation: You can use this method to simulate creation of a Personal Data record. During the simulation all necessary consistency checks are performed, but the record is not written to the database.

Availability: R/3 4.0A

Business Object: EmployeePersonalData

Documentation: As the name suggests this object stores personal data such as date of birth, place of birth, nationality, marital status, etc.

Application Area: Personnel Management

BAPI: EmployeePersonalData.Change

Documentation: This method can be used to change an entry.

Availability: R/3 4.0A

BAPI: EmployeePersonalData.Create

Documentation: This method creates a new entry.

Availability: R/3 4.0A

BAPI: EmployeePersonalData.CreateSuccessor

Documentation: This method creates a new entry and copies all non-specified data fields from the previous entry.

Availability: R/3 4.0A

BAPI: EmployeePersonalData.Delete

Documentation: This method deletes an entry.

Availability: R/3 4.0A

BAPI: EmployeePersonalData.GetDetail

Documentation: This method shows all information stored for one entry.

Availability: R/3 4.0A

BAPI: EmployeePersonalData.GetList

Documentation: This method retrieves the all entries of an employee.

Availability: R/3 4.0A

BAPI: EmployeePersonalData.Request

Documentation: This method creates a new non-active entry, which can be approved later.

Availability: R/3 4.0A

BAPI: EmployeeInternalControl.Simulatecreation

Documentation: You can use this method to simulate creation of a Personal Data record. During the simulation all necessary consistency checks are performed, but the record is not written to the database.

Availability: R/3 4.0A

Business Object: EmployeePrivAddress

Documentation: Documentation currently not available

Application Area: Personnel Management

BAPI: EmployeePrivAddress.Approve

Documentation: You can use this method to unlock an **Addresses** record. It enables you to implement the "double verification principle" in conjunction with the REQUEST method. According to this principle, at least two users must take part in writing an active record to the database. One user creates a locked record using the REQUEST method. The other user unlocks or ' activates' this record using the APPROVE method. password.

Availability: R/3 3.0E

BAPI: EmployeePrivAddress.Change

Documentation: You can use this method to change an **Addresses** record.

Availability: R/3 3.0E

BAPI: EmployeePrivAddress.Create

Documentation: You can use this method to create an **Addresses** record.

Availability: R/3 3.0E

BAPI: EmployeePrivAddress.Createsuccessor

Documentation: You can use this method to create a subsequent **Addresses** record.

Availability: R/3 3.0E

BAPI: EmployeePrivAddress.Delete

Documentation: You can use this method to delete an **Addresses** record.

Availability: R/3 3.0E

BAPI: EmployeePrivAddress.Getdetail

Documentation: You can use this method to read an **Addresses** record.

Availability: R/3 3.0E

BAPI: EmployeePrivAddress.GetList

Documentation: You can use this method to read the keys (instances) of **Addresses** records which are within the specified time interval.

Availability: R/3 3.0E

BAPI: EmployeePrivAddress.Request

Documentation: You can use this method to create a locked **Addresses** record. It enables you to implement the "double verification principle" in conjunction with the APPROVE method. According to this principle, at least two users must take part in writing an active record to the database. One user creates a locked record using the REQUEST method. The other user unlocks or ' activates' this record using the APPROVE method.

Availability: R/3 3.0E

BAPI: EmployeePrivAddress.Simulatecreation

Documentation: You can use this method to simulate creation of an Addresses record. During the simulation all necessary consistency checks are performed, but the record is not written to the database.

Availability: R/3 3.0E

Business Object: EmployeeTrip

Documentation: The business object Employee trip contains the employee business trips. A trip represents an employee' s change of location, including the journeys to and from the destination, so that he or she can temporarily carry out tasks away from his or her usual place of work. A trip contains several trip facts, which consist mainly of data on trip times, trip destinations, and trip expenses. The data is used to calculate reimbursement amounts. Each trip has exactly one status which characterizes the respective stage of its "life cycle": applied for; carried out; accounted, etc. Individual specifications concerning the allocation of costs to cost centers, projects, manufacturing orders, etc. can be made for a trip or for the individual trip facts.

Application Area: Financial Accounting

BAPI: EmployeeTrip.Approve

Documentation: This method approves an employee trip.

Availability: R/3 4.0A

BAPI: EmployeeTrip.Cancel

Documentation: This method cancels an employee trip.

Availability: R/3 4.0A

BAPI: EmployeeTrip.CollectMileage

Documentation: This method returns the business mileage of an certain employee.

Availability: R/3 4.0A

BAPI: EmployeeTrip.CreateFromData

Documentation: This method is used to create an intance of an business trip for a specific employee.

Availability: R/3 4.0A

BAPI: EmployeeTrip.CreateFromDataWeekly

Documentation: This method is used to create an instance of a weekly report for an employee.

Availability: R/3 4.0A

BAPI: EmployeeTrip.Delete

Documentation: This method deletes an employee trip.

Availability: R/3 4.0A

BAPI: EmployeeTrip.ExistenceCheck

Documentation: This method checks whether a specific business trip is known.

Availability: R/3 4.0A

BAPI: EmployeeTrip.GetDetails

Documentation: This method enables the user to access further information about a specific business trip of an employee.

Availability: R/3 4.0A

BAPI: EmployeeTrip.GetDetailsWeekly

Documentation: This method enables the user to access further information about a specific business trip of an employee on a weekly basis.

Availability: R/3 4.0A

BAPI: EmployeeTrip.GetExpenseForm

Documentation: Using the EmployeeTrip.GetExpenseForm method, the trip costs standard form is returned. At the present time only the return of single trips is supported within the object. If an error occurs while the method is performed, this is indicated in the return parameter.

Availability: R/3 3.1G

BAPI: EmployeeTrip.GetExpenseForm_HTML

Documentation: This method is identical to EmployeeTrip.GetExpenseForm except for the output is provided as an HTML document.

Availability: R/3 3.1G

BAPI: EmployeeTrip.GetList

Documentation: Using the method EmployeeTrip.GetList all of an employee' s trips within a specific time interval are retrieved. The method returns the value RETURN, which contains an exact error message in the case of errors. This method also provides the amounts for the trips, if desired.

Availability: R/3 3.1G

BAPI: EmployeeTrip.GetOptions

Documentation: This business object returns relevant data on a specific employee and business trip.

Availability: R/3 4.0A

BAPI: EmployeeTrip.GetOptionsWeekly

Documentation: No documentation available.

Availability: R/3 3.1G

BAPI: EmployeeTrip.GetStatus

Documentation: Using the method EmployeeTrip.GetStatus, the status of a single trip is returned. The amounts of the trip are also provided. The method returns the value RETURN, which contains an exact error message in the case of errors.

Availability: R/3 3.1G

BAPI: EmployeeTrip.SetOnHold

Documentation: This method sets the status of an business trip on hold.

Availability: R/3 4.0A

Business Object: FixedAsset

Documentation: The business object Fixed asset is an object, a right or another item of the enterprise' s assets which is intended for long-term use, and can be individually identified in the balance sheet. The development of the values of fixed assets can be viewed as a whole, or broken down into the component parts (sub-numbers) of assets.

Application Area: Financial Accounting

BAPI: FixedAsset.GetDetail

Documentation: This method retrieves the most important master data for afixed asset in Asset Accounting. With this method, data which is valid on the report date is selected from the time-dependent data for an asset (for example cost center). You can enter the report date as a parameter when you call up the method. If you do not specify a report date, the method uses the current date.

Availability: R/3 3.1G

Business Object: FunctionalArea

Documentation: The business object Functional area contains the organizational unit formed for the purposes of accounting that subdivides the ENTERPRISE according to the requirements of cost-of-sales accounting - the FUNCTIONAL AREA. The FUNCTIONAL AREA corresponds to a task area in the ENTERPRISE to which costs of sales can be allocated.

Application Area: Financial Accounting

BAPI: FunctionalArea.ExistenceCheck

Documentation: This method ascertains if the functional area specified exists in the system. The result of the check is recorded in a return code.

Availability: R/3 3.1G

BAPI: FunctionalArea.GetDetail

Documentation: This method provides more information about a functional area. It records information determined by the system and a return code.

Availability: R/3 4.0A

BAPI: FunctionalArea.GetList

Documentation: This method lists existing functional areas and instructions for them.

Availability: R/3 4.0A

Business Object: GeneralLedger

Documentation: The business object General ledger is a ledger that is defined for the representation of values in general ledger accounting that are used as a basis for preparation of the balance sheet as well as the profit and loss statement.

Application Area: Financial Accounting

BAPI: GeneralLedger.GetGLAccountBalance

Documentation: Documentation currently not available

Availability: R/3 4.0A

BAPI: GeneralLedger.GetGLAccountPeriodBalances

Documentation: Documentation currently not available

Availability: R/3 4.0A

BAPI: GeneralLedger.GetGLAccountCurrentBalance

Documentation: Documentation currently not available

Availability: R/3 4.0A

Business Object: GeneralLedgerAccount

Documentation: The business object General ledger account comprises the General ledger account transaction figures. It is a structure for the recording of value movements in a company code that relate to a value category that is specified by a chart of accounts item.

Application Area: Financial Accounting

BAPI: GeneralLedgerAccount.ExistenceCheck

Documentation: This method enables you to establish whether a general ledger account exists for a given company code. The result of the check is denoted by a return code.

Availability: R/3 3.1G

BAPI: GeneralLedgerAccount.GetBalance

Documentation: This method returns the final balance for a general ledger account for a specified fiscal year. The user can choose whether the balance is returned in the transaction currency of the company code currency. The exceptions here are those general ledger accounts kept in different transaction currencies where the balance can be given in the company code currency only. The currency and the balance as determined by the system are returned, with any problems arising being reported in the form of a return code message.

Availability: R/3 3.1G

BAPI: GeneralLedgerAccount.GetCurrentBalance

Documentation: This module returns the closing balance for a G/L account for the current fiscal year. The user can choose whether the balance is displayed in the transaction or company code currency (with the exception of G/L accounts that are managed in different transaction currencies). For such accounts, only the company code currency can be used. The balance determined by the system together with the currency are returned, with problems arising being output as a return code.

Availability: R/3 3.1G

BAPI: GeneralLedgerAccount.GetDetail

Documentation: This method provides the user with further information about a general ledger account. Detail data determined by the system concerning the general ledger account is returned. Problems arising are returned in the form of a return code message.

Availability: R/3 3.1G

BAPI: GeneralLedgerAccount.GetList

Documentation: This method provides a list of general ledger accounts for a company code. The data determined by the system is returned in a table. Problems arising are returned in the form of a return code message.

Availability: R/3 3.1G

BAPI: GeneralLedgerAccount.GetPeriodBalances

Documentation: This method returns the G/L account balances per posting period for a given fiscal year. The user can choose whether a balance is returned in the transaction or company code currency except for G/L accounts managed in different transaction currencies. Only the company code currency is returned for such accounts. The balance as determined by the system and the currency are returned with any problems appearing in the form of a return code.

Availability: R/3 3.1G

Business Object: Helpvalues

Documentation: Documentation currently not available

Application Area: Cross-Application Components

BAPI: Helpvalues.GetList

Documentation: This method returns the permitted input values for a field which you pass in a BAPI call. The input values are stored in corresponding check tables in the R/3 System. To determine the input values, this method uses the help view linked to the field in the ABAP/4 Dictionary. As the result depends on the field, it is returned in two unstructured tables:

Availability: R/3 3.1H

Business Object: InspectionLot

Documentation: The business object inspection lot is a request to a plant to carry out a quality inspection for a specific quantity of material.

Structure:

Inspection lots can contain several operations. An operation describes the activity to be carried out at a work center.

Operations can contain several inspection characteristics. An inspection characteristic describes what must be inspected. Inspection specifications such as the sample size and the number of samples, as well as the conditions under which the entire lot quantity should be accepted or rejected are predefined for each inspection characteristic.

Inspection operations can also contain one or more inspection points.

Inspection lots for customer deliveries, production orders, or run schedule headers can be divided into several partial lots.

Comments:
Inspection lots can be created for goods movements or deliveries to customers. They can also be created for production orders or run schedule headers, so the material can be inspected during production. The inspection results recorded for an inspection lot provide the basis for making the usage decision.

Application Area: Quality Management

BAPI: InspectionLot.GetList

Documentation: This method generates a list of inspection lots for given selection criteria. Inspection lots can be selected on the basis of a material, batch, order, run schedule header, vendor, or customer.

Availability: R/3 4.0A

BAPI: InspectionLot.GetOperations

Documentation: This method generates a list of operations for a given inspection lot.

Availability: R/3 4.0A

BAPI: InspectionLot.StatInterface

Documentation: This method transfers the results for quantitative inspection characteristics to an external statistical system and triggers the statistical analyses.

Availability: R/3 4.0A

Business Object: InspLotCharacter

Documentation: The business object inspection characteristic is subordinate to an inspection operation. An inspection characteristic defines what must be inspected. Inspection specifications such as the sample size and the number of samples, as well as the conditions under which the entire lot quantity should be accepted or rejected are predefined for each inspection characteristic.

Application Area: Quality Management

BAPI: InspLotCharacter.GetRequirements

Documentation: This method reads the inspection requirements for a given inspection lot characteristic.

Availability: R/3 4.0A

BAPI: InspLotCharacter.GetResult

Documentation: This method reads the inspection results for a given inspection lot characteristic. The method provides the following details:

- Inspection results on characteristic level (with status, mean value, and so on)
- Inspection results on sample level
- Single inspection results

Availability: R/3 4.0A

BAPI: InspLotCharacter.SetResult

Documentation: This method creates new inspection results for a given inspection lot characteristic. It is also possible to update the existing results. In this case, the results can be read with InspLotCharacter.GetResult. Both methods use the same data structures.

Availability: R/3 4.0A

Business Object: InspLotOperation

Documentation: The object inspection operation defines all activities for inspection characteristics using specific test equipment at a specific work center.

Application Area: Quality Management

BAPI: InspLotOperation.GetList

Documentation: This method generates a list of operations for a given inspection lot.

Availability: R/3 4.0A

BAPI: InspLotOperation.GetChar

Documentation: This method generates a list of characteristics for a given inspection lot operation. With a filter (processing mode), you can limit the selection on the basis of the status of the characteristic.

Availability: R/3 4.0A

Business Object: InspPoint

Documentation: The object inspection point is a reference point for carrying out several inpections within an operation.

Application Area: Quality Management

BAPI: InspPoint.GetList

Documentation: This method generates a list of inspection points for a given inspection lot operation. If many inspection points exist, you can limit the selection by defining lower and upper limits for the inspection point number.

Availability: R/3 4.0A

BAPI: InspPoint.CreateFromData

Documentation: This method creates a new inspection point for a given inspection lot operation. The data contains the inspection point identification and optionally, the partial lot, batch and inspection point valuation. The inspection point number is assigned internally and is supplied as an export parameter.

Availability: R/3 4.0A

BAPI: InspPoint.Change

Documentation: This method changes the partial lot, batch, and valuation of an inspection point. The inspection point identification cannot be changed. The method uses the same data structure as InspPoint.GetList and InspPoint.GetFromData.

Availability: R/3 4.0A

BAPI: InspPoint.GetRequirements

Documentation: This method generates the inspection requirements for an inspection point. The method supplies the following data:

- Catalog data for valuating an inspection point
- Information that specifies whether the inspection point is based on time or quantity
- Description of fields for the inspection point identification
- Various flags for the data required to confirm an inspection point

Availability: R/3 4.0A

Business Object: InternalOrder

Documentation: The business object Internal order contains orders that are created for cost monitoring, and, if required, operating result monitoring purposes. A controlling order can be an instrument for the monitoring of values in connection with individual measures (for example, repairs) or services to be rendered. Furthermore, it can also be used to monitor values to be controlled separately, for example, the overhead costs of an organizational unit.

Application Area: Controlling

BAPI: InternalOrder.GetDetail

Documentation: This method supplies the most important master data, system status, user status, and business transactions allowed for the object "Internal order". Investment orders or revenue-recording orders are not retrieved.

Availability: R/3 3.1G

BAPI: InternalOrder.GetList

Documentation: This method selects internal orders using various criteria, except for investment or revenue-recording orders. The following information is returned:

- Number of orders found
- All orders found

You can combine selection criteria as desired. However, you must include limiting requirements; you cannot select all internal orders simultaneously.

Availability: R/3 3.1G

Business Object: InvestmentProgram

Documentation: The business object InvestmentProgram is a hierarchical structure representing business plans for producing fixed assets or providing services during an approval year.

A InvestmentProgram consists of a number of positions. Investment program positions can group together several controlling objects, for which the capital investment program carries the budget and plan values.

Application Area: Investment Management

BAPI: InvestmentProgram.ExistenceCheck

Documentation: You can use this method to check whether a capital investment program exists.

Availability: R/3 4.0A

BAPI: InvestmentProgram.GetLeaves

Documentation: You can use this method to determine all the leaf positions of a given capital investment program, that is, all positions of the program hierarchy without any subordinate positions.

Availability: R/3 4.0A

BAPI: InvestmentProgram.GetRequestsAndLeaves

Documentation: You can use this method to determine all appropriation requests which are assigned to the leaf positions of a given capital investment program. In addition, the method also returns all leaf positions of the investment program.

Availability: R/3 4.0A

Business Object: Kanban

Documentation: Documentation currently not available

Application Area: Production Planning and Control

BAPI: Kanban.GetListForSupplie1

Documentation: This method reads the relevant kanban information from the customer' s system for a vendor who delivers material to his customers using KANBAN methods.

Availability: R/3 4.0A

BAPI: Kanban.GetListForSupplier

Documentation: This method reads the relevant KANBAN information from the customer' s system for a supplier who delivers goods to his/her customer using KANBAN product control.

Availability: R/3 3.1G

BAPI: Kanban.SetInProcess

Documentation: This method sets the kanbans in the KANBAN table to the status IN PROCESS.

Availability: R/3 3.1G

Business Object: Location

Documentation: Documentation currently not available

Application Area: Training and Event Management

BAPI: Location.GetListAll

Documentation: You can use this method to display all available business event locations.

Availability: R/3 3.1G

Business Object: Material

Documentation: The Business object Material is a tangible or intangible good that is the subject of business activity. A material is traded, used in manufacture, consumed, or produced. In order that data relating to the material can be stored at the relevant organizational level, a material master record is structured in accordance with the organizational units plant, storage location, warehouse complex, and distribution chain. Different units of measurement can apply to a material. A conversion factor allowing conversion into the base unit is defined for each such unit. A material can have an International Article Number (EAN) for identification purposes, depending on the unit of measurement. Different parameters define how activities such as materials planning or quality inspection are carried out for a material. An individual piece of a material can be distinguished from others by a serial number. Materials made up of several constituent parts can be described with the aid of bills of material (BOMs). Materials for which a change in requirements has occurred are recorded in the planning file for the purpose of requirements planning. The different methods by which a material can be produced are determined in a production version. Subsets of a material that are manufactured in a particular production run and stocked separately from other subsets of the same material are managed as batches. The batch where-used list indicates which batches make up a certain finished product or which finished products contain a certain batch.

Application Area: Logistics-General

BAPI: Material.Availability

Documentation: Using this method, you can determine the receipt quantity still available for a particular material in a certain plant according to Availability-To-Promise logic (MRPII). The availability check is carried out on transferring the material number and the plant. The scope of the check, that is, which stocks, receipts and issues are to be included in the check is defined by the combination of checking group (material master) and checking rule. In the method, the system uses the checking rule defined in Sales & Distribution (A). The results of the availability contain dates and available receipt quantities (ATP quantities). The results of the check depend on the following entries:

- If no date and no quantity is transferred, the system retrieves the ATP situation from today' s date into the future as the result.
- If only a date and no quantity is transferred, the system retrieves the ATP situation from the corresponding date as the result.

- If both a date and a quantity are transferred, the system calculates the availability situation for the quantity specified.

The system also retrieves the end of the replenishment lead time.

Availability: R/3 3.1G

BAPI: Material.ExistenceCheck

Documentation: Documentation currently not available

Availability: R/3 3.1G

BAPI: Material.GetBatchCertificate

Documentation: This method creates a ready-made quality certificate for a batch of the material. Currently, only the PDF format (Adobe Acrobat(TM) Reader) is supported.

Availability: R/3 3.1G

BAPI: Material.GetInternalNumber

Documentation: Documentation currently not available

Availability: R/3 4.0A

BAPI: Material.GetList

Documentation: Documentation currently not available

Availability: R/3 4.0A

BAPI: Material.GetBatches

Documentation: This method creates a list of batches for a material. You can limit this list by using the entry parameters. For example, you can create a list of batches for a material, whose expiration date has not been exceeded.

Availability: R/3 3.1G

BAPI: Material.GetDetail

Documentation: This method provides detailed data for a specified material.

Availability: R/3 3.1G

Business Object: MaterialReservation

Documentation: Documentation currently not available

Application Area: Materials Management

BAPI: MaterialReservation.CreateFromData

Documentation: This method creates a new material reservation.

Availability: R/3 4.0A

BAPI: MaterialReservation.GetItems

Documentation: This method displays reservations.

Availability: R/3 4.0A

BAPI: MaterialReservation.GetDetail

Documentation: This method shows all information stored for one reservation.

Availability: R/3 4.0A

Business Object: Network

Documentation: The business object Network describes the flow of activities of a project to be implemented, that is, how and in what order the tasks required for implementation of a project are to be carried out. A network consists of activities that in turn can be subdivided by activity elements. The activities of a network are arranged in a net pattern on the basis of technical, content-related or time-related dependencies between the activities. Activities and activity elements can represent different aspects of the processing. It is possible to distinguish accordingly between activities and activity elements that are processed internally and those that are processed externally. Independently of the type of processing, there are also activities and activity elements that represent cost aspects solely. To activities can be assigned production tools that are required for their performance. The tasks described in activities or activity elements can be explained in more detail in standardized project information.

Application Area: Project System

BAPI: Network.ExistenceCheck

Documentation: This method checks whether a specific network is known.

Availability: R/3 4.0A

BAPI: Network.GetDetail

Documentation: This method enables the user to access further information about a network with all its subparts such as header milestone and components.

Availability: R/3 4.0A

BAPI: Network.GetInfo

Documentation: You can use this method to read detailed information abot networks, including all objects, from the system.

Availability: R/3 4.0A

BAPI: Network.Maintain

Documentation: This method allows the user maintain of objects, retrieved by the GetDetail method. The method includes the handling of the system status release, lock and unlock.

Availability: R/3 4.0A

Business Object: OpenInfoWarehouse

Documentation: The Documentation is currently not available.

Application Area: Cross-Application Components

BAPI: OpenInfoWarehouse.GetCatalog

Documentation: This method can be used to read the meta data of the Open Information Warehouse (OIW). This method provides the information objects, data sources and the information objects for each of the data sources.

Availability: R/3 3.1G

BAPI: OpenInfoWarehouse.GetData

Documentation: This method allows data to be read from the data sources of the Open Information Warehouse (OIW). You can restrict the scope of the data read by defining selection criteria and by specifying the number of columns (that is, information objects of the OIW metadata) you require.

Availability:R/3 3.1G

Business Object: PieceOfEquipment

Documentation: The business object Piece of equipment contains the piece of equipment as an individual physical object that is to be maintained separately. A piece of equipment can be, for example, a machine, an auxiliary resource, testing or measuring equipment, or customer equipment. Since a piece of equipment can be used in different ways in the course of time, it has, in addition to its main features, time-dependent features (for example, location data, planning responsibilities or account assignments) that are recorded in the piece of equipment - utilization. A piece of equipment can itself be composed of other pieces of equipment. This is represented by means of the equipment hierarchy. Through the cross-linking of pieces of equipment, the environment of a piece of equipment becomes clear. This is important, for example, in order to recognize, when analyzing malfunctions, possible causes in upstream pieces of equipment, or to show, when planning maintenance tasks, which upstream or downstream pieces of equipment must be switched off if necessary. To a piece of equipment different maintenance partners can be assigned in dependence upon their respective partner role.

Application Area: Plant Maintenance

BAPI: PieceOfEquipment.CreateFromData

Documentation: With this method, you create a piece of equipment. The data for the piece of equipment to be created includes:

- The master data of the piece of equipment
- The usage periods of the piece of equipment
- The short texts of the piece of equipment
- The location data of the piece of equipment
- The sales data of the piece of equipment
- This method also assigns equipment numbers, if desired.

Availability: R/3 3.1H

BAPI: PieceOfEquipment.DismantleAtFuncloc

Documentation: You use this method to dismantle a piece of equipment at a functional location at a specified time. If the dismantling is performed successfully, the method returns the updated equipment data.

Availability: R/3 4.0A

BAPI: PieceOfEquipment.DismantleFromHierarchy

Documentation: You use this method to dismantle a piece of equipment from an equipment hierarchy at a specified time. If the dismantling is performed successfully, the method returns the updated equipment data.

Availability: R/3 4.0A

BAPI: PieceOfEquipment.GetCatalogProfile

Documentation: With this method, you determine the catalog profile for the current date for a piece of equipment. If the method can not determine a catalog profile for the piece of equipment itself, for its construction type, for the functional location at which the piece of equipment is installed or for its construction type, the catalog profile that is assigned to the notification type is used.

Availability: R/3 3.1G

BAPI: PieceOfEquipment.GetDetail

Documentation: This method reads details of a piece of equipment. The piece of equipment can be identified either by the equipment number or by a combination of material number and serial number. If the piece of equipment specified exists and you have authorization to retrieve the data, the method delivers details about:

- The master data of the piece of equipment
- The usage periods of the piece of equipment
- The short texts of the piece of equipment
- The location data of the piece of equipment
- The sales data of the piece of equipment
- The hierarchy data of the piece of equipment

Availability: R/3 3.1G

BAPI: PieceOfEquipment.GetListForCustomer

Documentation: This method returns a list of all pieces of equipment for a customer.

Availability: R/3 4.0A

BAPI: PieceOfEquipment.InstallAtFuncloc

Documentation: You can use this method to install a piece of equipment at a functional location at a specified time. If the installation is performed successfully, the method returns the updated equipment data.

Availability: R/3 4.0A

BAPI: PieceOfEquipment.InstallInHierarchy

Documentation: You can use this method to install a piece of equipment in an equipment hierarchy at a specified time. If the equipment has been successfully installed, the method returns the updated equipment data.

Availability: R/3 4.0A

BAPI: PieceOfEquipment.Update

Documentation: With this method, you change the data of a piece of equipment.

Availability: R/3 3.1H

Business Object: PlannedIndepReqmt

Documentation: The business object Planned independent requirement contains the requirement quantities of products planned for an anonymous market. A planned independent requirement specifies the entire requirement quantity of a plant material in dependence upon the independent requirement type and the independent requirement version. A planned independent requirement - schedule line is a subdivision of a planned independent requirement according to requirement quantity and date.

Application Area: Production Planning and Control

BAPI: PlannedIndepReqmt.Change

Documentation: Documentation currently not available

Availability: R/3 4.0A

BAPI: PlannedIndepReqmt.CreateFromData

Documentation: Documentation currently not available

Availability: R/3 4.0A

BAPI: PlannedIndepReqmt.GetDetail

Documentation: You can use this method to display the sales orders for a customer.

Availability: R/3 4.0A

Business Object: ProcurementOperation

Documentation Documentation currently not available

Application Area: Materials Management

BAPI: ProcurementOperation.GetCatalogs

Documentation: Documentation currently not available

Availability: R/3 4.0A

BAPI: ProcurementOperation.GetInfo

Documentation: Documentation currently not available

Availability: R/3 4.0A

Business Object: ProductCatalog

Documentation: The business object Product catalog is a means of presentation used in advertising grouping together advertising messages about a number of materials. A product catalog can occur in different variants, depending on the language and the currency in each case.

Application Area: Logistics-General

BAPI: ProductCatalog.GetItems

Documentation: This method returns items for a product catalog.

Availability: R/3 3.1H

BAPI: ProductCatalog.GetLayout

Documentation: This method can be used to read information for the layout of a product catalog. The layout defines the structure of a product catalog and contains information on where the materials are roughly placed in the piece.

Availability: R/3 3.1H

BAPI: ProductCatalog.GetLayoutDescription

Documentation: This method returns long texts for a layout area or a layout area item of a product catalog.

Availability: R/3 3.1H

BAPI: ProductCatalog.GetLayoutDocuments

Documentation: This method returns documents assigned to a layout area or layout area item in a product catalog.

Availability: R/3 3.1H

BAPI: ProductCatalog.GetList

Documentation: This method returns a list of all product catalogs belonging to the client. In addition to product catalog numbers, all short texts maintained (all languages) are also determined.

Availability: R/3 3.1G

BAPI: ProductCatalog.GetPrices

Documentation: This method returns prices for items in a product catalog.

Availability: R/3 3.1H

BAPI: ProductCatalog.GetSalesArea

Documentation: This method returns the sales area for an product catalog.

Availability: R/3 3.1H

BAPI: ProductCatalog.GetVariants

Documentation: This method returns variants for a product catalog. Short texts, languages and currencies of the variants are also determined.

Availability: R/3 3.1H

BAPI: ProductCatalog.GetDetail

Documentation: This method supplies header data for a product catalog.

Availability: R/3 4.0A

Business Object: ProfitCenter

Documentation: The business object Profit center contains organizational units formed for the purposes of accounting that subdivide the enterprise in a management-oriented manner, that is, for the purpose of internal control. For a profit center operating results can be shown that are determined according to cost-of-sales accounting and/or the "total cost" type of accounting. Through showing of the restricted assets the profit center can be extended to form the investment center. Profitability analysis at profit center level is based on costs and sales revenues. These are statistically assigned through parallel updating of all logical transactions relevant for a profit center and other allocations. The valuation of the deliveries and services performed between the profit centers can correspond with the valuation base in accounting or deviate from this base.

Application Area: Enterprise Controlling

BAPI: ProfitCenter.GetDetail

Documentation: This method supplies detailed information on the profit center master record.

Availability: R/3 3.1G

BAPI: ProfitCenter.GetList

Documentation: This method supplies a list of profit centers sorted by controlling area and person in charge within each controlling area.

Availability: R/3 3.1G

Business Object: ProjectDefinition

Documentation: The business object Project definition is a corporate project with a fixed objective that is to be achieved with set amounts of money, with the planned resources, and to an agreed level of quality in a given time. A corporate project is distinguished by its uniqueness, the risk it entails, and its importance within the organization. In the work breakdown structure it is structured according to the hierarchy of tasks and in the (Business Object) Network according to the flow of activities.

Application Area: Project System

BAPI: ProjectDefinition.ExistenceCheck

Documentation: This method checks whether a specific project definition is known.

Availability: R/3 4.0A

BAPI: ProjectDefinition.GetDetail

Documentation: This method enables the user to access further information about a project definition. The system returns both the system and user status, aswell as information on the project structure and longtext.

Availability: R/3 4.0A

BAPI: ProjectDefinition.CreateFromData

Documentation: You can use this method to create a project definition that contains data binding for the whole project. In the process, all the consistency checks are made that are made if you create a project definition using the R/3 transaction.

Availability: R/3 4.0A

BAPI: ProjectDefinition.Update

Documentation: You use this method to change a project definition and all the data that is binding for the whole project. The same consistency checks take place, that would do so if the changes were being made in the R/3 transaction.

Availability: R/3 4.0A

Business Object: PurchaseOrder

Documentation: The business object Purchase order is a request or instruction from a purchasing organization to a vendor (external supplier) or a plant to deliver a certain quantity of material or to perform certain services at a certain point in time. A purchase order (PO) consists of a number of items, for each of which a procurement type is defined. The following procurement types exist:

- Standard
- Subcontracting
- Consignment

- Stock transfer
- Service

The total quantity of the material to be supplied or the service to be performed set out in the PO item can be subdivided into partial quantities with corresponding delivery dates in the lines of a delivery schedule. With respect to PO items involving subcontracting work, the material components to be provided to the subcontractor can be specified for each delivery date. An item of the procurement type "service" comprises a set of service specifications. The specifications are structured by means of outline levels and the quantity ordered is set out in service lines. Value limits are stipulated in place of service lines for services that cannot be specified precisely. If services are to be released (ordered) against existing contracts, the PO item can contain a value limit relating to a certain contract.

Conditions can apply at various levels:

- to the entire purchase order
- at item level for a material to be supplied or the
- specifications for services
- at the level of the service line for an individual service

The costs incurred can be apportioned among different Controlling objects through the account assignment. The vendor can indicate to the purchasing organization that the delivery date will be met, or advise of any likely deviations therefrom, by sending various types of confirmation. The business transactions following on from the purchase order are documented item by item in the PO history.

Comment: Instead of the vendor as order recipient, other business partners can appear in various partner roles (as goods supplier or invoicing party, for example). In the purchase order, each item is destined for a plant. Each plant belongs to a company code, to which the creditor' s invoice is addressed.

Application Area: Materials Management

BAPI: PurchaseOrder.CreateFromData

Documentation: Envoking this method creates a purchase order.

Availability: R/3 4.0A

BAPI: PurchaseOrder.GetDetail

Documentation: This method displays detailed data of a purchase order, such as order type, purchasing group, release group, etc.

Availability: R/3 4.0A

BAPI: PurchaseOrder.GetItems

Documentation: This method displays line items of a purchase order.

Availability: R/3 4.0A

BAPI: PurchaseOrder.GetList

Documentation: You can use this method to list all purchase orders that have to be released (approved) with a certain release code and group (collective release).

Availability: R/3 4.0A

BAPI: PurchaseOrder.GetReleaseInfo

Documentation: This method enables you to display release information.

Availability: R/3 4.0A

BAPI: PurchaseOrder.Release

Documentation: You can use this method to release (approve) purchase orders. The PO number and the release code must be passed on. The new release status and the new release indicator are returned.

Availability: R/3 4.0A

BAPI: PurchaseOrder.ResetRelease

Documentation: You can use this method to cancel or revoke (reset) already effected releases of purchase orders. The PO number and the release code must be passed on. The release status and release indicator valid prior to the release are returned.

Availability: R/3 4.0A

Business Object: PurchaseReqItem

Documentation: Documentation currently not available

Application Area: Materials Management

BAPI: PurchaseReqItem.CreateFromData

Documentation: Method of creating a requirement coverage request.

Availability: R/3 3.1G

BAPI: PurchaseReqItem.GetList

Documentation: This method allows you to generate the following: All requirement coverage requests (purchase requisitions and reservations) created by a certain user. All purchase requisitions that have to be released with a certain release code and group (collective release) As of Release 4.0A, use the methods GetItems and GetItemsForRelease of the business object PurchaseRequisition.

Availability: R/3 3.1G

BAPI: PurchaseReqItem.Release

Documentation: This method enables you to release (signify approval of) purchase requisitions. This is necessary if a purchase requisition item satisfies the conditions defined in a release strategy in Customizing.

Availability: R/3 3.1G

BAPI: PurchaseReqItem.ResetRelease

Documentation: This method can be used to cancel the release (= revoke approval) of purchase requisitions. The requisition number/item and the release code must be passed on. The release status and the release indicator that were valid before the release are then returned.

Availability: R/3 3.1G

BAPI: PurchaseReqItem.SingleReleaeNoDialg

Documentation: This method enables you to release (signify approval of) purchase requisitions. This is necessary if a purchase requisition item satisfies the conditions defined in a release strategy in Customizing.

Availability: R/3 3.1G

Business Object: PurchaseRequisition

Documentation: The business object Purchase requisition contains the request or instruction to Purchasing to procure a certain quantity of a material or a service so that it is available at a certain point in time. A purchase requisition consists of a number of items, for each of which a certain procurement type is defined. The following procurement types exist:

- Standard
- Subcontracting
- Consignment
- Stock transfer
- Service

The item contains the quantity and delivery date of the material to be supplied or the quantity of the service to be performed. With respect to items involving subcontracting work, the material components to be provided to the subcontractor can be specified for each delivery date. An item of the procurement type "service" comprises a set of service specifications. The specifications are structured by means of outline levels and the quantity ordered and delivery date are set out in service lines. Value limits are stipulated in place of service lines for services that cannot be specified precisely. If services are requested against contracts, the requisition item can contain a value limit relating to a certain contract. The costs incurred as a result of the procurement can be apportioned among different Controlling objects through the account assignment.

Comment: A purchase requisition can be fulfilled by means of purchase orders or outline purchase agreements.

Application Area: Materials Management

BAPI: PurchaseRequisition.CreateFromData

Documentation: This method allows the creation a purchase requisition.

Availability: R/3 4.0A

BAPI: PurchaseRequisition.Change

Documentation: This method allows the user to change a purchase requisition.

Availability: R/3 4.0A

BAPI: PurchaseRequisition.Delete

Documentation: This method allows the user to delete a purchase requisition.

Availability: R/3 4.0A

BAPI: PurchaseRequisition.GetDetail

Documentation: This method enables the user to access further information about a purchase requisition such as order type, purchasing group, etc.

Availability: R/3 4.0A

BAPI: PurchaseRequisition.GetItems

Documentation: This method returns the line items of purchase requisition.

Availability: R/3 4.0A

BAPI: PurchaseRequisition.GetReleaseInfo

Documentation: This method enables you to display release information.

Availability: R/3 4.0A

Business Object: PurchasingInfo

Documentation: The business object Purchasing information is a source of information on the procurement of a certain material from a certain vendor. The purchasing information is subdivided by purchasing organization, allowing each purchasing organization to store details of the conditions it has negotiated and other procurement control data relating to the vendor concerned. There are purchasing informations for the following procurement types:

- Standard
- Pipeline

- Subcontracting

A purchasing info record can apply to the following organizational levels:

- purchasing organization
- plant

The different prices charged by a vendor for a material are logged in the order price history. Comment: The purchasing information enables Purchasing to obtain data such as the following at any time:

- Which materials a certain vendor has quoted for or supplied to date
- Which vendors have quoted for or supplied a certain material

Application Area: Materials Management

BAPI: PurchasingInfo.GetList

Documentation: To display purchasing info record.

Availability: R/3 4.0A

Business Object: QualityNotification

Documentation: The business object QualityNotification describes an object' s nonconformance with a quality requirement and contains a request to take appropriate action. A quality notification can take the form of a customer complaint, a complaint against a vendor, or an internal problem report.

Application Area: Quality Management

BAPI: QualityNotification.CreateFromData

Documentation: This method creates a quality notification. Using this function you can create set fields and the long text for the notification header, together with various notification items. Currently, you can only create notifications for notification types assigned to the original customer complaint .

Availability: R/3 3.1G

BAPI: QualityNotification.GetKeyFigures

Documentation: This method generates a list of related notifications or key figures for related notifications. This information can be used for statistical purposes.

Availability: R/3 4.0A

BAPI: QualityNotification.GetCatalogProfile

Documentation: This method selects the relevant Catalog profile for a combination of notification type and material number and displays the corresponding code groups and codes in a table. This method first checks whether a catalog profile is stored at the material level. If no profile is stored, the method selects the catalog profile assigned to the Notification type.

Availability: R/3 4.0A

BAPI: QualityNotification.GetListForCustomer

Documentation: Documentation currently not available

Availability: R/3 4.0A

BAPI: QualityNotification.GetMaterialListFCust

Documentation: Documentation currently not available

Availability: R/3 4.0A

Business Object: RetailMaterial

Documentation: Documentation currently not available

Application Area: Logistics-General

BAPI: RetailMaterial.Availability

Documentation: Using this function module, you can determine the receipt quantity still available for a particular material in a certain plant according to ATP logic (MRPII).

Availability:R/3 3.1G

BAPI: RetailMaterial.Clone

Documentation: You can use this method to create new material master data or to change existing material master data.

Availability: R/3 4.0A

BAPI: RetailMaterial.ExistenceCheck

Documentation: Documentation currently not available

Availability:R/3 3.1G

BAPI: RetailMaterial.GetBatchCertificate

Documentation: This method creates a table in MIME format. This table contains a quality certificate for a batch in edited form. Only the PDF (Adobe Acrobat™ Reader) format is currently supported. As well as the batch ID, you need to enter a customer number since the data in the quality certificate comes from a customer-specific view.

Availability: R/3 3.1G

BAPI: RetailMaterial.GetBatches

Documentation: This method is used to create a list of batches for a material. You can restrict this list by specifying input parameters. For example, you can create a list of the batches whose shelf life expiration date is not yet exceeded.

Availability: R/3 3.1G

BAPI: RetailMaterial.GetCharacsMerchandiseHierachy

Documentation: Documentation currently not available

Availability: R/3 4.0A

BAPI: RetailMaterial.GetComponents

Documentation: Documentation currently not available

Availability: R/3 4.0A

BAPI: RetailMaterial.GetDetail

Documentation: This method provides a selection of detail data for a given material. The data is valid for the whole client.

Availability: R/3 3.1G

BAPI: RetailMaterial.GetInternalNumber

Documentation: Documentation currently not available

Availability: R/3 4.0A

BAPI: RetailMaterial.GetList

Documentation: Documentation currently not available

Availability: R/3 3.1G

BAPI: RetailMaterial.GetVariantNumbers

Documentation: Documentation currently not available

Availability: R/3 4.0A

Business Object: SalesOrder

Documentation: The business object sales order is a contractual arrangement between a sales organization and a sold-to party concerning goods to be delivered or services to be rendered. A sales order contains information about prices, quantities and dates. The request is received by a sales area, which is then responsible for fulfilling the order. A sales order consists of several items that contain the quantity of the material or service specified for the order. This total quantity can be divided into different partial quantities with the corresponding delivery dates in the schedule lines. sales order items can be set in a hierarchy. This makes it possible to differentiate between batches or to seperate the different components of a material (BOM - bills of material). In the item conditions, the valid conditions for the items are listed. They can be distributed among the items from an overall condition valid for the whole order. One item can be divided into several billing plan dates that each determine a date for which the amount in the item is to be billed.

Application Area: Sales and Distribution

BAPI: SalesOrder.CreateFromDat1

Documentation: You can use this method to create sales orders. The system checks the order type because only orders with sales document category C are permitted.

Availability: R/3 4.0A

BAPI: SalesOrder.CreateFromData

Documentation: You can use this method to create sales orders. You have to enter at least the header data (using the ORDER_HEADER_IN structure) as well as partner data (using the ORDER_PARTNERS table) as the input parameters. After successfully creating the sales order, you receive both the document number and the detailed data about the partners involved. Errors that may occur are reported in the ERRORTEXT return parameter. The method provides detailed information about price and availability of an item.

Availability: R/3 3.1H

BAPI: SalesOrder.GetList

Documentation: With this method, you can retrieve the sales orders for a customer. In every case you must enter the required customer number and the required sales organization. The selection can be restricted by specification of certain criteria. For example, you can retrieve orders for a particular material or orders from a particular only. It is also possible to find a particular sales order via the purchase number of the customer.

Availability: R/3 3.1H

BAPI: SalesOrder.GetStatus

Documentation: You can use this method to find information for certain sales orders with regard to availability, processing status (for example, delivery status) and prices.

Availability: R/3 3.1H

BAPI: SalesOrder.Simulate

Documentation: This method has the same interface definition as the BAPI_SALESORDER_CREATEFROMDATA function but differs from it in that here the sales order is not updated. Here you can determine availability and pricing. This data is displayed in the ORDER_ITEMS_OUT table.

Availability: R/3 3.1H

Business Object: ServiceEntrySheet

Documentation: Documentation currently not available

Application Area: Materials Management

BAPI: ServiceEntrySheet.GetDetail

Documentation: Documentation currently not available

Availability: R/3 4.0A

BAPI: ServiceEntrySheet.GetList

Documentation: Documentation currently not available

Availability: R/3 4.0A

BAPI: ServiceEntrySheet.GetReleaseInfo

Documentation: You can use this method to display the sales orders for a customer.

Availability: R/3 4.0A

BAPI: ServiceEntrySheet.Release

Documentation: Documentation currently not available

Availability: R/3 4.0A

BAPI: ServiceEntrySheet.ResetRelease

Documentation: Documentation currently not available

Availability:R/3 4.0A

Business Object: ServiceNotification

Documentation: The business object Service notification is the description of a service-relevant situation that has occurred and, if required, the request to react to this situation. A service notification can be a customer notification, a service request or a service activity notification. A customer notification describes a malfunction in a service object at a customer' s and requests its repair. A service request describes a service activity desired by the customer and requests that this be carried out. This service activity does not relate to a malfunction. A service activity notification describes service activities that have been carried out and that were not based on a customer notification or a service request. A service notification consists of several items that contain a description of the damage that has occurred at a customer' s or of the service activity requested or performed. For a service notification item there can also be a cause description and a detailed description of the activity already performed. In a service notification task can be specified for the service notification additional tasks that are to be planned and documented and whose status is to be followed. To a service notification can be assigned different partners in dependence upon their respective partner role.

Application Area: Plant Maintenance

BAPI: ServiceNotification.CreateFromData

Documentation: You can use this method to create service notifications. This allows you to create certain fields of the notification header, the long text for the notification header and several notification items.

Availability: R/3 3.1E

BAPI: ServiceNotification.GetList

Documentation: This method selects service notifications assigned to a customer. For this, the following data is used as selection criteria: the customer that is maintained in the service notification the notification date whereupon all notifications are selected, whose date is greater than or equal to the data entered. You can choose to limit the selection using a partner function and partner number. The service notifications are returned. All notifications are selected regardless of status. You can choose to exclude completed notifications from the selection.

Availability: R/3 4.0A

Business Object: TimeAvailSchedule

Documentation: Documentation currently not available

Application Area: Personnel Time Management

BAPI: TimeAvailSchedule.Build

Documentation: This method determines employees' availability on the basis of their personal work schedule and any time data that represents an exception to the schedule. An employee' s personal work schedule is determined from the long-term work schedule in the Planned Working Time infotype (2003), and from the short- and medium-term changes to the planned working time in the Substitutions infotype (2003). This gives a valid daily work schedule for each day containing detailed specifications on planned working times, break times, and overtime. An employee' s availability can depend on these planned times, or on exceptions to the daily work schedules in the Absences (2001), Attendances (2002), and Overtime (2005) infotypes. A list of time points provides information on an employee' s availability. Each entry indicates the valid availability information from one time point (date, time) to the next time point. The first entry for a personnel number is always for 00:00 on the start date of the selection period, and the last entry for 24:00 on the end date.

Availability: R/3 4.0A

Business Object: TimeMgtConfirmation

Documentation: Documentation currently not available

Application Area: Personnel Time Management

BAPI: TimeMgtConfirmation.Post

Documentation: This method allows confirmations from Logistics to be transferred to Human Resources. Confirmations that you want to transfer as work time events are saved in the interface table EVHR. Confirmations that you want to transfer as durations are saved in the interface table LSHR.

Availability: R/3 4.0A

Business Object: WorkBreakdownStruct

Documentation: The business object Work breakdown structure contains the work breakdown structure element and the work breakdown structure element - hierarchy. The structure of a project specified by a project definition is described in the work breakdown structure. This work breakdown structure represents a hierarchical structuring of the project in different work breakdown structure elements, whereby the structuring is generally multi-level. A work breakdown structure element can be a specific task that describes an activity necessary to achieve the objective connected with the project. The work breakdown structure forms the basis for organization and coordination in the project. On the basis of the work breakdown structure elements of a work breakdown structure the costs of a project and the related expenditure of work and time can be planned and the accumulated expenditure of work and time as well as the accrued costs can be shown in the course of project implementation. Work breakdown structure elements can be explained in standardized project information.

Application Area: Project System

BAPI: WorkBreakdownStruct.GetInfo

Documentation: You use this method to read detailed information about project definitions and the WBS elements of certain projects from the system.

Availability: R/3 4.0A

BAPI: WorkBreakdownStruct.Maintain

Documentation: This method provides the following functionality:

- Create PSP elements (priority 1)
- Change PSP elements (priority 1)
- Release PSP elements (priority 1)
- Lock PSP elements (priority 1)
- Unlock PSP elements (priority 1)
- Set deletion flags (priority 1)
- Remove deletion flags (priority 1)
- Change dates of PSP elements (priority 1)
- Change hierarchy (priority 1)
- Set milestones (priority 2)
- Change milestones (priority 2)
- Delete milestones (priority 2)
- Create long texts (priority 2)

Change long texts (priority 2)

Availability: R/3 4.0A

Appendix B: Glossary

ABAP/4

Advanced Business Application Programming/4. ABAP/4 is SAP's fourth generation programming language. ABAP/4 is used to develop dialog applications and to evaluate databases, e.g. via reports.

ABAP/4 Repository

All of SAP R/3's development objects the ABAP/4 Development Workbench are stored in the so-called ABAP/4 Repository. The ABAP/4 development objects include: ABAP/4 programs, screens, documentation etc.

Application Link Enabling (ALE)

ALE is R/3's Middleware that enables implementation of networks of multiple SAP systems. ALE provides the links for integrating business processes, avoiding rigid ties to a central database supported by predefined business rules. This is also known as a 'loosely coupled but tightly integrated' system. It involves the exchange of consistent master, control or transactional data on loosely linked R/3 modules. To achieve tight integration, synchronous and asynchronous communication is used.

Application Programming Interface (API)

An API supports the developer with a set of routines that are used by a program when carrying out services for instance by the operating system or other applications.

Attributes

Attributes contain data about Business Objects, thus describing a particular object property.

Business Application Programming Interfaces (BAPIs)

BAPIs are methods for accessing R/3's Business Objects. Based on SAP's Remote Function Call (RFC) Technology external software is able to link directly to R/3. BAPIs are stable and reusable interfaces, which enable developers in using Java, Visual Basic, Lotus Script and other languages to develop new application components for the R/3 environment that will be usable over R/3 Version changes.

Business Component

Behind the term Business Component is nothing new. With the introduction of SAP's Business Framework Architecture the name for a R/3 module a.k.a. R/3 application was changed into Business Component (e.g. HR, SD, MM...)

Business Framework Architecture (BFA)
The Business Framework Architecture is an open, integrated component-based architecture that includes R/3 applications and third-party products and technology. With SAP's Business Framework, customers are now able to install new R/3 Business Components without the need of upgrading the entire R/3 System. With SAP's R/3 Release 3.1, the Business Framework Architecture was introduced.

Business Navigator
The Business Navigator is the R/3 tool that lists the implemented R/3 Reference Model.

Business Object
The Business Object is a key R/3 System Object. Data models are assigned to Business Objects in the R/3 Repository. A Business Object is a representative of a specific business entity (e.g. Employee) in a R/3 Business Components (e.g. HR-Module).

Business Object Repository (BOR)
All SAP Business Objects types and their methods are identified and described in the R/3 Business Object Repository. The BOR contains all meta-data about the SAP Business Objects.

Intermediate Document (IDoc)
An IDoc is a data-container, which is used within an ALE scenario to send data from one R/3 system to another R/3 system. An IDoc has the following structure: A control record, several data records and various status records.

Key Fields
The Key Fields of a Business Object determine the structure of an identifying key, which allows an application to access a specific instance of the object type. The object type "Material" and the key field "Material.Number" can be seen as examples of an object type and its corresponding key field.

R/3 Reference Model
The R/3 Reference Model represents the functional capability of the R/3 System. The R/3 Reference Model is a graphical representation of all aspects of a companies business. For instance information flows, data and organisation structures, task sequences in a chronological order, etc. The R/3 Reference Model is based on best practices, which helps a R/3 implementation as a base for modelling and eventually improving the customer's business processes.

R/3 Automation Software Development Kit
The R/3 Automation Software Development Kit also know as RFCSDK contains available RFC libraries and RFC DLLs (Dynamic

Link Libraries), dialogs, error handling functions and documentation.

Remote Function Call (RFC)

A Remote Function Fall, which is SAP's implementation of Remote Procedure Call (RPC) in ABAP/4, is a call to a function module that runs on a different system as the calling function. The remote function call can also be called from within the same system but usually the caller and callee will be dispersed.

Transactional Remote Function Call (tRFC)

Beginning from R/3 Release 3.0, data can be transferred between two R/3 systems reliably and safely using transactional RFCs (tRFC). The exact execution of one function module is carried out using a unique Transaction Identifier (TID), which is exchanged between the Client and the R/3 Server program, and ensures that the called function module is executed exactly once in the RFC server system.

SAP Business Workflow

SAP Business Workflow links information and business objects together with processes across transactions. R/3 has a built-in relationship between its object model and workflow. The workflow itself consists of a sequence of work items that is handled by the workflow manager and controlled by R/3's event related response mechanism. The main task of SAP's workflow management is to support the information flow and to reduce the workload of R/3 users. It offers services that can simplify and speed up the completion of companies business processes.

Bibliography

[BG96] Rüdiger Buck-Emden, Jürgen Galimow: *Die Client/Server-Technologie des SAP-Systems R/3*. Addison-Wesley, 1996. ISBN 3-8273-1021-0

[Boo94] Grady Booch: *Object-Oriented Analysis and Design (Second Edition)*. The Benjamin/Cummings Publishing Company, Inc., 1994. ISBN 0-8053-5340-2

[Bro97] A. Barry, T. Brown: *Researching Groupware use in situ, Work in Progress*. Department of Sociology, University of Surrey, Guildford (GB), 1997

[Cla97] Christiane Clasen: Im Mix aus Groupware und Intranet liegt der Vorteil. In *Computerwoche*, Page: 47-48, Nr. 20, 1997

[CB94] William R. Cheswick, Steven M. Bellovin: *Firewalls & Internet Security, Repelling the Wily Hacker*. Addison-Wesley Professional Computing Series, 1994. ISBN 0-201-63357-4

[Date95] C. J. Date: *An introduction to database systems*. Addison-Wesley, 6th Edition, 1995. ISBN 0-201-82458-2

[DHLW96] Carl Dudley, Khawar Hameed, Dr. Kecheng Liu, Sue Williams: Lecture Notes: *Database Systems*. School of Computing, Staffordshire University, UK, 1996.

[EGR91] C. Ellis, S. Gibbs, G. Rein: Groupware, Some Issues and Experiences. In *Communication of the ACM*, 34(1):38-58, January 1991

[Ell93] Clarence Ellis: Formal and Informal model of office activity. In *Proceedings of Information Processing 93, the 9th IFIP World Computer Congress*, Paris, 1993

[Eng93] H. Engesser: *Duden Informatik,* Edition 2. Dudenverlag, 1993. ISBN 3-411-05232-5

[EW94] Clarence Ellis, Jacques Wainer: Computer-Supported Co-operative Work. In *Proceedings of ACM CSCW' 94 Conference on Computer-Supported Co-operative Work,* page 79-88, Association for Computing Machinery 1994

[Forr96] Forrester Research, Inc. *Forrester Software Strategy Report, E-Mail Shootout.* June 1996.

[Fitz96] Wolfgang Fitznar: *SAP R/3-Einführung, Grundlagen, Anwendungen, Bedienung für Release 3,* CDI (Hrsg.) Markt&Technik-Verlag, 1996. ISBN 3-8272-5096-X

[GS91] Greenberg, Saul (Hrsg.): *Computer-supported Co-operative Work and Groupware,* London, Academic Press Ltd, 1991

[Holl94] David Hollingsworth: The Workflow Reference Model. In *The Workflow Management Coalition Specification.* The Workflow Management Coalition, Brussels, 1994

[Hunt93] Craig Hunt: *TCP/IP Network Administration.* O'Reilly & Associates, Inc., 1993

[IBM93] *IBM Dictionary of Computing.* McGraw-Hill, 1993.

[IBM96] *Lotus Notes Release 4.5. A Developer's Handbook.* International Business Machines Corporation, 1996.

[Joo94] Stef Joosten: An empirical study about the practice of workflow management. In *Workflow Analysis in 12 different organisations. WA-12 Report,* University of Twente, Enschede, The Netherlands, 1994

[KW96] Rüdiger Kretschmer, Wolfgang Weiss: *SAP-R/3-Entwicklung mit ABAP/4*. SYBEX, 1996. ISBN 3-8155-7232-0

[KL89] Kim Won, Frederick H. Lochovsky: *Object-Oriented Concepts, Databases and Applications,* ACM Press 1989

[Lotus96a] Lotus Development Corporation: Lotus Communiqué: *Lotus Notes and SAP R/3: Complementary Solutions Working Together*. A Lotus Development Corporation Strategy Brief, November 1996.

[Lotus96b] Lotus Development Corporation: White Paper: *Lotus Notes Enterprise and DBMS Integration*. August 1996.

[Lotus97] *The LotusScript Extension for SAP R/3*. Lotus Development Corporation, 1997

[Mai95] Dr. Maier: Lecture Notes: *Marketing*. Fachhochschule Konstanz, 1995.

[Muth97] Michael Muthig: *Im- und Export-Schnittstellen des integrierten Systems SAP R/3*. Diplomarbeit an der Fachhochschule Konstanz, Betreuer: Prof. Dr. Ralf Leibscher, 1997.

[MW96] John Middleton, Euan Willson: Lecture Notes: *System development through packages*. School of Computing, Staffordshire University, UK, 1996.

[PW96a] Price Waterhouse, *Technology Forecast 1996 5.4 Groupware, Imaging, and Document Management, Page 489-506*

[PW96b] Price Waterhouse, *Technology Forecast 1996 6.2 Corporate Applications, Page 577-592*

[PW96c] Price Waterhouse, *Technology Forecast 1996* 4.3 *Security, Page 203-240*

[PW97a] Price Waterhouse, *Technology Forecast 1997 5.4 Groupware, Workflow and Document Management, Page 343-366*

[PW97b] Price Waterhouse, *Technology Forecast 1997 6.2 Corporate Applications, Page 397-422*

[Rich97] Cate Richards, Jane Calabria, Rob Kirkland, David Hatter, Roy Rumaner, Susan Trost, Tim Vallely, Mark Williams: *Special Edition. Using Lotus Notes and Domino 4.5.* QUE, 1997. ISBN 0-7897-0943-0

[SAP96a] *Functions in Detail, R/3 System, SAP Business Workflow.* SAP AG, 1996

[SAP96b] The SAP R/3 System in the World Wide Web, Netting the benefits. In *SAPInfo,* No. 49/50 – 6/96. SAP AG, 1996.

[SAP96c] *R/3 System, The Business Framework.* SAP AG, September 1996

[SAP96d] *Secure Network Communications – Integrating R/3 into Network Security Products.* SAP AG, August 1996.

[SAP96e] *Functions in Detail, R/3 System, Technological Infrastructure.* SAP AG, September 1996.

[SAP96f] *Functions in Detail, R/3 System, SAP Business Workflow.* SAP AG, September 1996.

[SAP96g] *The Foundation for Genuine – Business on the Internet.* SAP AG, October 1996

[SAP96h] *R/3 integrated with network security products.* In SAP Info D&T No. 51 – 8/96. Page: 27 - 28. SAP AG, 1996

[SAP96i] *R/3 System, SAP Business Objects.* SAP AG, September 1996

[SAP96j] SAP Documentation: *System R/3, the ABAP/4 Development Workbench, Remote Communications.* R/3 System Release 3.0.
SAP AG, 1996.

[SAP96k] *ALE – Application Link Enabling.* SAP AG, Mai 1996.

[SAP97a] From Client/Server to Internet Architecture. In *R/3 System, SAP Technology Infrastructure,* 1997.

[SAP97b] *R/3 System Release 3.1G, Online Documentation.* SAP AG, March 1997.

[SAP97c] *SAP- Business Application Programming Interface (BAPI) – Overview.* Dr. Jörg Wiederspohn, Silvia Gruben, Andreas Hirche. PACA90, SAP AG, July 1997.

[SAP97d] *BAPI Introduction and Overview.* Version R/3 Release 3.1 H. SAP AG, July 1997.

[SAP97e] *R/3 System Release 3.1G Online Documentation.* SAP AG, March 1997

[SAP97f] *BAPI Programming.Version R/3 Release 4.0.* SAP AG, December 1997

[SAP98] Marketing Department, SAP AG, November 1998

[Ste96] Dominik Stein: *Definition und Klassifikation der Begriffswelt um CSCW, Workgroup Computing, Groupware, Workflow Management.* Universität Gesamthochschule Essen, 1996

[Stro97] Ulrich Strobel-Vogt, Prof. Dr. Paul Wenzel [Hrsg.]: *SAP Business Workflow in der Logistik.* Vieweg, 1997. ISBN 3-528-05599-5

[TSMB95] Computerunterstützung für die Gruppenarbeit. Addison-Wesley, 1995. ISBN 3-89319-878-4

[Tan89] Andrew S. Tanenbaum: *Computer Networks, Second Edition.* Prentice Hall International Editions, 1989. ISBN 0-13-166836-6

[WHSH96] Liane Will, Christiane Hienger, Frank Straßenburg, Rocco Himmer: *R/3 Administration.* Addison-Wesley, 1996. ISBN 3-89319-967-5

[Wil88] Paul Wilson. Editor: R. Speth: Key Research in Computer Supported Co-operative Work (CSCW). In *Proceedings of the EUTECO' 88 Conference*, page 211-226, Vienna, 1988

[WH97] John-Harry Wieken, Thomas Hoffmann: Objektorientiertes Design ohne Geheimnis. Heise, 1997. ISBN 3-88229-074-9

[@DEC] DEC LinkWorks
http://www.digital.com/info/linkworks

[@IBM] IBM FlowMark
http://www.software.ibm.com/ad/flowmark/

[@IDC] International Data Corporation (IDC)
http://www.idcresearch.com

[@InConcert] InConcert, Inc.
http://www.inconcertsw.com

[@Lotus] Lotus Development Corporation
http://www.lotus.com

[@OMG] Object Management Group
http://www.omg.org/

[@SAP] SAP AG
http://www.sap-ag.de/ or http://www.sap.com

[@Siebel] Siebel Systems, Inc.
http://www.siebel.com/

[@WFMC] Workflow Management Coalition
http://www.aiai.ed.ac.uk/project/wfmc/

Index

T

V

W

X